Fundamentals of Quantum Entanglement

IOP Series in Coherent Sources and Applications

About the Editor

F J Duarte is a laser physicist based in Western New York, USA. He has a 30+ year experience in the academic, industrial and defense sectors. Duarte is editor/author of 13 laser optics books and sole author of two books (*Tunable Laser Optics* and *Quantum Optics for Engineers*). He has made original contributions to the field of narrow-linewidth tunable laser oscillators, organic laser gain media, nanoparticle solid-state laser materials, and laser interferometry. He is also the author of the multiple-prism grating dispersion theory applicable to tunable lasers, laser pulse compression, and coherent microscopy. Duarte is a Fellow of the Australian Institute of Physics (1987) and a Fellow of the Optical Society of America (1993). He has been awarded the Paul F Foreman Engineering Excellence Award and the David Richardson Medal from the Optical Society.

Coherent Sources and Applications

Since the discovery of the laser, applications of this wondrous emitter of coherent radiation have grown enormously. Subsequently, we have also become familiar with additional sources of coherent radiation such as the free electron laser, optical parametric oscillators, and interferometric emitters. The aim of this new book series is to explore and explain the physics and technology of widely applied sources of coherent radiation and to match them with utilitarian and cutting-edge scientific applications. Selected coherent sources are those that offer advantages in particular emission characteristics areas such as broad tunability, high spectral coherence, high energy, or high power. An additional area of inclusion is those coherent sources capable of high performance in the miniaturized realm. Selected uses include practical applications valuable to the industrial, commercial, and medical sectors. Particular attention will be given to scientific applications with a bright future scope such as coherent (or laser) spectroscopy, astronomy, biophotonics, space communications, space interferometry, and quantum entanglement.

Publishing benefits

Authors are encouraged to take advantage of the features made possible by electronic publication to enhance the reader experience through the use of colour, animation and video, and incorporating supplementary files in their work.

Do you have an idea of a book that you'd like to explore?

For further information and details of submitting book proposals, see iopscience. org/books or contact Ashley Gasque at ashley.gasque@iop.org.

Fundamentals of Quantum Entanglement

F J Duarte

IOP Publishing, Bristol, UK

ISBN   978-0-7503-2228-7 (ebook)
ISBN   978-0-7503-2226-3 (print)
ISBN   978-0-7503-2227-0 (mobi)

DOI   10.1088/2053-2563/ab2b33

Version: 20191001

IOP Expanding Physics
ISSN 2053-2563 (online)
ISSN 2054-7315 (print)

British Library Cataloguing-in-Publication Data: A catalogue record for this book is available from the British Library.

Published by IOP Publishing, wholly owned by The Institute of Physics, London

IOP Publishing, Temple Circus, Temple Way, Bristol, BS1 6HG, UK

US Office: IOP Publishing, Inc., 190 North Independence Mall West, Suite 601, Philadelphia, PA 19106, USA

A mis nietos.

Contents

Appendices

Preface

Fundamentals of Quantum Entanglement is a monograph, including 29 short chapters and 10 appendices, dedicated to explain the origin and meaning of the probability amplitude for quantum entanglement $|\psi\rangle = (|x_1, y_2\rangle - |y_1, x_2\rangle)$. This is the probability amplitude that describes the polarization entanglement of two quanta ($n = 2$) propagating in two different directions ($N = 2$). The initial approach is from a chronological perspective and explains both the philosophical path, initiated by Einstein, Podolsky, and Rosen (EPR), and the alternative physics path, initiated by Dirac, Wheeler, Pryce, and Ward, to what eventually became the field of quantum entanglement. As the subject matter progresses, attention is paid to the hidden variable theories that led to the formulation of Bell's theorem that highlights the incompatibility of such theories with quantum mechanics. Experiments designed to test for these 'hidden theories' in light of Bell's inequalities are also described. Attention is then focused on a clear and transparent interferometric derivation of the ubiquitous quantum entanglement probability amplitude for $n = N = 2$, $|\psi\rangle = (|x_1, y_2\rangle - |y_1, x_2\rangle)$, and on the extension of this interferometric framework to include probability amplitudes applicable to $n = N = 2^1, 2^2, 2^3 \ldots 2^r$ and $n = N = 3, 6, 9$ situations. This direct non-philosophical interferometric approach, *à la Dirac*, to quantum entanglement removes much of the mystery associated with this subject matter. Following a description of applications of the quantum entanglement probability amplitude to cryptography, teleportation, and quantum computing, the focus returns to the essence of quantum entanglement and matters of interpretation. In this regard, the pragmatic approach to quantum mechanics associated with Dirac, Feynman, Lamb, and Ward is reinforced.

Acknowledgments

The genesis for this monograph first originated from a number of discussions, in the 1998–2000 period, between the author and John Clive Ward. The idea gained further impetus following a series of lectures on quantum mechanics the author delivered at the State University of New York. Throughout the intervening years, and while working on *Quantum Optics for Engineers*, the author maintained sporadic and yet stimulating discussions on the general challenges facing quantum communications with Thomas M Shay, then at the US Air Force. In a more recent setting, the author has had the fortune of discussing various non-classified aspects of space-to-space quantum entanglement communications with Travis S Taylor, from the US Army. Support from *Interferometric Optics*, during the gestation of this work, is gratefully acknowledged.

Author biography

F J Duarte

F J Duarte is a Chilean-born laser physicist based in Western New York, USA. He graduated with First Class Honours in physics from Macquarie University where he was a student of the well-known quantum physicist J C Ward. At Macquarie, he also completed his PhD research on optically-pumped molecular lasers in 1981. He then became a Postdoctoral Fellow at the University of New South Wales, where he built UV narrow-linewidth tunable lasers for high-resolution IR–UV double-resonance spectroscopy. Duarte has worked and contributed professionally in the academic, industrial, and defense sectors, and has practiced physics in Australia, the Americas, and Europe. He is the author of numerous refereed papers and an inventor in the fields of lasers and optics. Duarte is the editor/author of 14 scholarly books, including *Dye Laser Principles*, *High-Power Dye Lasers*, *Organic Lasers and Organic Photonics*, *Tunable Laser Applications*, and *Tunable Lasers Handbook*. He is sole author of *Tunable Laser Optics* and *Quantum Optics for Engineers*. Duarte has made original contributions in the fields of coherent imaging, high-power tunable lasers, laser metrology, liquid and solid-state organic gain media, narrow-linewidth tunable laser oscillators, organic semiconductor coherent emission, N-slit quantum interferometry, polarization rotation, quantum entanglement, and space-to-space secure interferometric communications. He is also the author of the generalized multiple-prism grating dispersion theory, and pioneered the use of Dirac's quantum notation in interferometry and classical optics. His contributions have found applications in astronomical instrumentation, dispersive optics, femtosecond laser microscopy, geodesics, gravitational lensing, laser isotope separation, laser medicine, laser pulse compression, laser spectroscopy, nonlinear optics, polarization optics, and tunable diode laser design. Current interests include semiconductor organic lasers, tunable laser physics, interferometric theory via Dirac's notation, and the foundations of quantum entanglement. Duarte is a Fellow of the Australian Institute of Physics (1987) and a Fellow the Optical Society of America (1993). He has received the *Engineering Excellence Award* (1995) and the *David Richardson Medal* (2016) from the Optical Society.

Fundamentals of Quantum Entanglement

F J Duarte

Chapter 1

Introduction

The experimental origin of quantum mechanics is highlighted, and essential features of the photon are described. The philosophical path to quantum entanglement, as defined by the works of EPR → Bohm and Aharonov → Bell, and the physics path to quantum entanglement, as defined by the works of Dirac → Wheeler → Pryce and Ward, are introduced and discussed. The probability amplitude for quantum entanglement $|\psi\rangle = (|x_1, y_2\rangle - |y_1, x_2\rangle)$, discovered in 1947, is then introduced prior to a high-level overview of the field. This introduction then concludes with a chapter-by-chapter description of *Fundamentals of Quantum Entanglement* and its purpose.

1.1 Introduction

Very few subjects in physics evoke interest in the general population like that generated by quantum entanglement. This interest is intrinsic, fueled by general curiosity on everything quantum, and is also stimulated by writers trying to explain quantum entanglement effects while using words such as *strange*, *mysterious*, and even *weird*. The aim of this monograph is to illuminate, and elucidate, the physics and the fundamentals of quantum entanglement to scientists and engineers with a background in first-year physics and first-year mathematics.

1.2 A few words on quantum mechanics

Quantum mechanics was discovered around 1900 by Max Planck. It was the result of an experimental discovery in the macroscopic domain. Planck was attempting to explain the energy distribution of light sources as a function of frequency. In this endeavor, he introduced, in the absence of derivation, one of the most momentous equations in the history of physics (Planck 1901):

$$E = h\nu \tag{1.1}$$

where E is the energy in units of Joules (J), ν is the frequency of light in units of Hz, or s^{-1}, and $h = 6.626\ 069\ 57 \times 10^{-34}$ J s is Planck's constant. This equation links

doi:10.1088/2053-2563/ab2b33ch1

high frequencies, or wavelengths toward the blue end of the spectrum, with high energies. Meanwhile, low frequencies, or wavelengths toward the red end of the spectrum, are associated with low energies.

There are several equivalent ways to approach quantum mechanics, the most prevalent being:

1. Heisenberg's matrix mechanics (Heisenberg 1925, Born and Jordan 1925, Born *et al* 1926);
2. Schrödinger's equation (Schrödinger 1926);
3. Dirac's *bra–ket* notation (Dirac 1939);
4. Moyal's statistics (Moyal 1949);
5. Feynman's path integrals (Feynman and Hibbs 1965).

The presentation in this monograph is entirely based on Dirac's *bra–ket* notation. This notation is explained in chapter 2.

1.2.1 The photon from a quantum perspective

The concept of a single photon, or *quantum*, is extraordinarily important to the field of quantum entanglement. Equally important are pairs of photons or pairs of *quanta*. In this subsection, relevant thoughts and reflections on the photon by three of the brightest luminaries in the field of quantum physics are included in an effort to provide a manifold of useful concepts on the photon:

- 'The wave function gives information about the probability of *one* photon being in a particular place and not the probable number of photons in that place .... Each photon then interferes only with itself. Interference between two different photons never occurs' (Dirac 1978).
- 'Newton thought that light was made up of particles, but then it was discovered that it behaves like a wave ... We say: "It is like *neither*"' (Feynman *et al* 1965).
- 'Photons cannot be localized in any meaningful manner, and they do not behave at all like particles, whether described by a wave function or not' (Lamb 1995).

In regard to Dirac's dictum on interference, it should be clarified that in quantum mechanics *two indistinguishable quanta are the same quantum*, or photon. Indistinguishable photons are photons of *the exact same frequency*.

The concept of *nonlocality* of the photon, brought up by Lamb, is of paramount importance to quantum optics and can be intuitive to experienced experimentalists: 'All the indistinguishable photons illuminate the array of N slits, or grating, simultaneously. If only one photon propagates, at any given time, then that individual photon illuminates the whole array of N slits simultaneously' (Duarte 2003). The property of nonlocality is highlighted because it is essential to quantum entanglement.

In more quantitative terms: a single photon moves, in vacuum, at the speed of light $c = 2.997\ 924\ 58\ \text{m s}^{-1}$. It has a wavelength λ, and a frequency ν, so that

$$\lambda = \frac{c}{\nu}. \tag{1.2}$$

As already mentioned, a photon is related to a quantum energy $E = h\nu$, or

$$E = \hbar\omega \tag{1.3}$$

where $\hbar = h/2\pi$ and $\omega = 2\pi\nu$. A single photon exhibits a quantum momentum of (de Broglie 1924)

$$p = \hbar k \tag{1.4}$$

where $k = 2\pi/\lambda$ is known as the wave number.

As mentioned by Dirac (1978), a single photon is associated with *complex wave functions* of the form

$$\psi(x, t) = \psi_0 e^{-i(\omega t - kx)}, \tag{1.5}$$

and these complex wave functions can be used to provide a mathematical representation of the probability amplitudes.

Furthermore, it can be established from Heisenberg's uncertainty principle (Dirac 1978)

$$\Delta p \Delta x \approx h \tag{1.6}$$

that single photons can be extremely nonlocal and can exhibit *enormous coherence lengths* as described by

$$\Delta \nu \Delta x \approx c. \tag{1.7}$$

The extreme nonlocality comes from the fact that a single quantum can exhibit an extraordinarily narrow linewidth $\Delta\nu$. As will be seen further in this monograph, this condition of nonlocality is central to quantum entanglement. Finally, 'ensembles of indistinguishable photons exhibiting very narrow linewidth $\Delta\nu$ originating from nearly monochromatic sources, such as narrow-linewidth lasers, approximate the behavior of a single photon' (Duarte 2014).

1.3 Ward's observation

'$(|x, y\rangle - |y, x\rangle)$ ... *was my first lesson in quantum mechanics, and in a very real sense my last, since all the rest is mere technique, which can be learnt from books*' (Ward 2004).

1.4 History of quantum entanglement

'*If I have seen further it is by standing on the sholders of giants,*' wrote Newton in 1675. This sentence is as relevant to physics now as it was then. It beautifully underscores the importance of history in physics. More specifically, it can be interpreted as a visionary call from the master to stimulate proper and meaningful referencing in the field of physics.

The vast majority of works published in the field of quantum entanglement present an incomplete version of history heavily biased toward the philosophical while neglecting the physics origin of quantum entanglement. Even though the philosophical perspective is extraordinarily enthralling, and has served quantum entanglement well by attracting the attention of the public at large, the less glamorous physics needs to be a participant in order to provide a complete and transparent exposition of the subject.

In this monograph, both the philosophical and the physics aspects of quantum entanglement are presented and discussed. This presentation is an extension of previous ideas and discussions on the subject (Dalitz and Duarte 2000, Duarte 2012, 2013a, 2013b, 2014, 2016).

1.4.1 The philosophical path

The philosophical path to quantum entanglement is well known to those familiar with the literature on the subject. It all started with a paper written by Einstein *et al* (1935) that became one of the most cited papers in physics. In this paper, which became known as EPR using the initials of the surnames of the three authors Einstein, Podolsky, and Rosen, questions were formulated on the completeness of quantum mechanics while introducing the notion of a 'possible' more complete formulation of the subject. At the same time, Einstein corresponded with Schrödinger, who also published on the subject, and introduced the word *entanglement* to describe the action at a distance introduced by the thought experiment considered by Einstein and colleagues (Schrödinger 1935, 1936). As it turns out, the EPR paper eventually generated an enormous citation following while the Schrödinger papers languished in the archives.

Key in deciding the chain of events that followed EPR was a paper by Bohm and Aharonov (1957) that discussed the EPR paper in the context of developments from the physics side of things, notably the work of Wu and Shaknov (1950), while also introducing a discussion on *hidden variables*. An additional key contribution was a famous paper by Bell (1964) that referenced Bohm and Aharonov while providing a transparent proof that quantum mechanics was incompatible with hidden variables. Incompatibility was determined by a probabilistic argument culminating in a probabilistic inequality. Violation of this inequality meant incompatibility of quantum mechanics with hidden variables. Here, it should be mentioned that hidden variable theories were though to provide a possible deterministic explanation of quantum experimental measurements.

In fairness to history, it should also be mentioned that the concept of incompatibility between quantum mechanics and hidden variables was introduced as early as 1932 by von Neumann; however, it was claimed that the proof leading to the von Neumann (1932) conclusion was found 'wanting' (Bell 1964).

Post-Bell came a paper by Clauser *et al* (1969) that made the use of Bell's theorem applicable to optical experiments and introduced modified Bell inequalities. Next came the use of those modified Bell inequalities by Aspect *et al* (1981, 1982a, 1982b) in optical experiments that demonstrated violation of Bell's inequalities. This chain of events is illustrated in figure 1.1.

Following the Aspect experiments, the citation trail of published papers in the field followed almost exclusively the philosophical path, which can be summarized as EPR → Bohm and Aharonov → Bell.

The label 'philosophical path' does not mean that it was all pure philosophy. There was also physics but the researchers were interested and motivated by a deeper meaning of quantum mechanics framed in a deterministic way of thinking compatible with classical notions. The fact that quantum mechanics denied that deeper understanding apparently made this philosophical avenue quite irresistible.

1.4.2 The physics path

The physics path is a pragmatic, measurement-driven, avenue to quantum entanglement. It began with a paper by Dirac on 'pair theory' (Dirac 1930) and some 16 years later was followed by a transparent and profound statement by John Wheeler that captures the essence of quantum entanglement: '*if one of these photons is linearly polarized in one plane, then the photon that goes off in the opposite direction with equal momentum is linearly polarized in the perpendicular plane*' (Wheeler 1946), our italics. Wheeler made his statement in reference to a positron–electron annihilation process that leads to the emission of two quanta in opposite directions, $e^+e^- \rightarrow \gamma_1\gamma_2$.

Wheeler's paper was followed by a publication that correctly predicted the quantum mechanical cross-section for γ ray scattering in experiments designed to test Wheeler's prediction (Pryce and Ward 1947). The quantum scattering equation

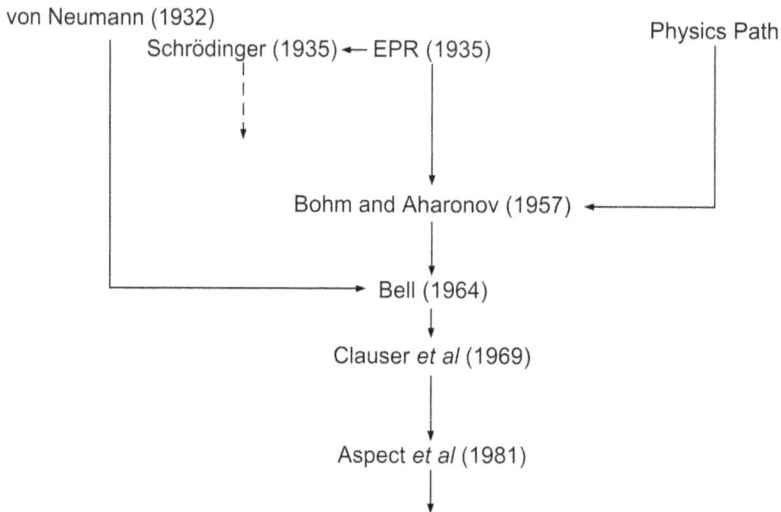

Figure 1.1. Philosophical path to quantum entanglement. Solid arrowed lines indicate a direct citation path in the literature. The main citation path runs vertically downward. For instance, von Neumann's contribution was cited by Bell but not necessarily by other authors in the direct vertical path. The papers by Schrödinger received intermittent citations for many years and only began to be cited regularly since the 1990s. The assigned years refer to literature emergence dates.

of Pryce and Ward was corroborated theoretically by Snyder *et al* (1948) and experimentally by Hanna (1948), Bleuler and Bradt (1948), and Wu and Shaknov (1950).

Following the publication of the $\sim\frac{1}{2}$ page disclosure entitled 'Angular correlation effects with annihilation radiation' (Pryce and Ward 1947), Ward followed with a disclosure of the derivation of the quantum entanglement probability amplitude, $|\psi\rangle = (|x_1, y_2\rangle - |y_1, x_2\rangle)$, as part of his dual-topic doctoral thesis (Ward 1949). It should be noted that this probability amplitude is essential to the derivation of the final correct quantum scattering equation published by Pryce and Ward (1947).

It should also be stated that $|\psi\rangle = (|x_1, y_2\rangle - |y_1, x_2\rangle)$ includes and contains all the physics relevant to quantum entanglement experiments. All this was done in a complete vacuum of philosophical discussions and in the total absence of concern of, or preoccupation with, hidden variable theories.

In the physics path to quantum entanglement, illustrated in figure 1.2, it is apparent that after the Wu and Shaknov (1950) experiments, Wu and coworkers revisited the entanglement arena this time utilizing cross-field fertilization from the philosophical path, namely with knowledge of Bell's theorem (Kasday *et al* 1975). As it turns out, practitioners in the philosophical path doubted the relevance of the Wu experiments from a local hidden variable perspective (Clauser *et al* 1969, Clauser and Horner 1974). Following this line of thought, Aspect *et al* (1981) wrote, 'the experiments agree with QM predictions. However, because of the lack of efficient polarizers for 0.5 MeV photons, strong supplementary assumptions are

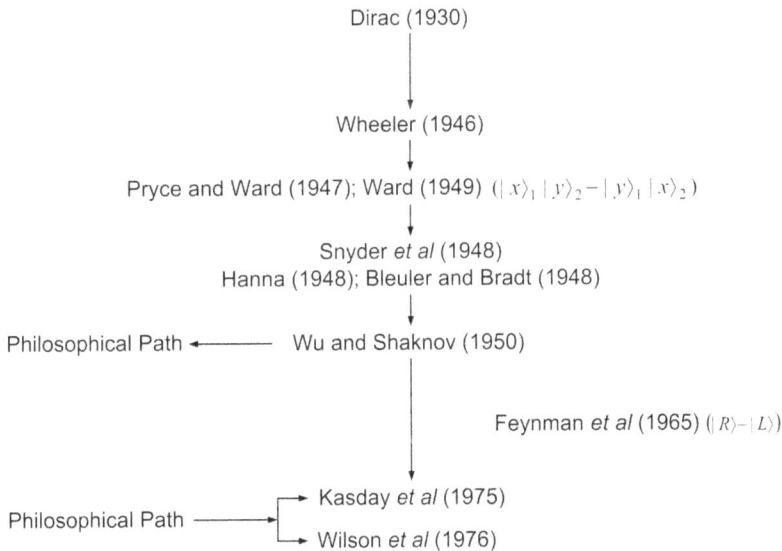

Dirac (1930)

$\downarrow$

Wheeler (1946)

$\downarrow$

Pryce and Ward (1947); Ward (1949) $(|x\rangle_1 |y\rangle_2 - |y\rangle_1 |x\rangle_2)$

$\downarrow$

Snyder *et al* (1948)
Hanna (1948); Bleuler and Bradt (1948)

$\downarrow$

Philosophical Path $\longleftarrow$ Wu and Shaknov (1950)

Feynman *et al* (1965) $(|R\rangle - |L\rangle)$

Philosophical Path $\longrightarrow$ Kasday *et al* (1975)
Wilson *et al* (1976)

Figure 1.2. Physics path to quantum entanglement. Solid arrowed lines indicate a direct citation path in the literature. The main citation path runs vertically downward. The paper by Wu and Shaknov (1950) influenced the paper by Bohm and Aharonov (1957) in the philosophical avenue while the papers by Kasday *et al* (1975) and Wilson *et al* (1976) were influenced by Bell's theorem. The assigned years refer to literature emergence dates.

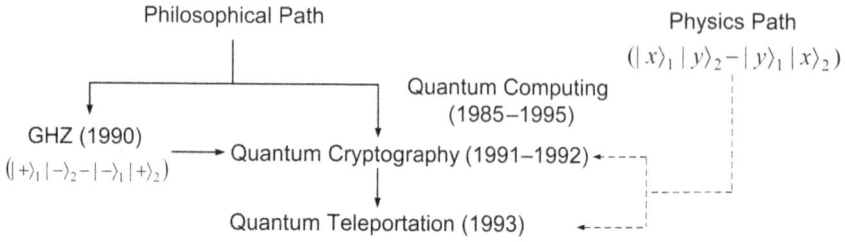

Figure 1.3. The literature from the philosophical path converges directly into quantum entanglement-based subfields such as quantum cryptography (and/or quantum communications) and quantum teleportation. The influence of the physics path is largely limited to the widespread use of $|\psi\rangle = (|x\rangle_1|y\rangle_2 - |y\rangle_1|x\rangle_2)$, in the absence of acknowledgments. Quantum computing emerged in 1985 via a Feynman contribution and from the paper that introduced the term *qbit* in 1995 (see text). A noticeable fraction of authors in contemporaneous quantum computing research do refer to EPR and Bell's theorem.

necessary to interpret these results via Bell's theorem'. This was despite the initial favorable reception given by Bohm and Aharonov (1957) to the experiment by Wu and Shaknov (1950).

What should be kept in mind is that even though today all of the developments in the field of quantum entanglement revolve around the probability amplitude for quantum entanglement, in its several versions, there is almost no acknowledgment of its origin or the physics path that led to its discovery. This monograph is designed to provide a perspective on quantum entanglement from the philosophical and the physics perspectives by including all the relevant literature. This approach should help remove the cloud of mystery that surrounds quantum entanglement.

In conjunction with figures 1.1 and 1.2, the overall perspective of the field is given in figure 1.3.

1.5 The field of quantum entanglement

The emergence of the combined words *quantum entanglement*, in the open literature, appears to go back to the mid to late 1980s (Ghirardi *et al* 1987). This was a few years after the optical experiments on quantum entanglement by Aspect *et al* (1981, 1982a, 1982b).

Today the field of quantum entanglement is divided roughly into three subfields as outlined in figure 1.3: quantum cryptography (which includes quantum communications), quantum teleportation, and quantum computing. On paper, judging by citations, these subfields have been heavily influenced by the ideas and concepts derived from the philosophical path to quantum entanglement.

Also on paper, and judging from citations, the acknowledgment of the physics path has been miniscule. This almost non-existing recognition has persisted albeit the all-important probability amplitude for quantum entanglement, which was discovered in a vacuum of philosophical arguments:

$$|\psi\rangle = \frac{1}{\sqrt{2}}(|x\rangle_1|y\rangle_2 - |y\rangle_1|x\rangle_2) \tag{1.8}$$

was reintroduced into the mainstream of quantum entanglement as

$$|\psi\rangle = \frac{1}{\sqrt{2}}(|+\rangle_1|-\rangle_2 - |-\rangle_1|+\rangle_2) \tag{1.9}$$

by Greenberger *et al* (1990) in the same paper that introduced the GHZ probability amplitude.

In other words, although equation (1.8) began to be used heavily in quantum cryptography and quantum teleportation, virtually no mention has been made of its origin back in 1947. This is quite a remarkable situation.

In figure 1.3, quantum cryptography is dated as having been initiated in 1991 since that is the date applicable to the introduction of entangled photon pairs to the field (Ekert 1991). However, quantum cryptography as a concept goes back to the early–mid 1980s (see, for example, Bennett and Brassard 1984). Quantum teleportation was introduced by Bennett *et al* (1993).

Some of the first discussions on quantum computing are due to Feynman (1985, 1986), and reflecting on Feynman's independence, his approach to quantum computing was his own and thus it did not follow a pre-established literature path. The term *qbit*, short for quantum bit, was introduced by Schumacher (1995). Schumacher's approach, although it does not mention the EPR $\rightarrow$ Bohm and Aharonov $\rightarrow$ Bell philosophical path, does touch on issues of quantum interpretation and thus it might also qualify as a step in the philosophical path. A subsequent paper by Steane (1998) deals directly with EPR–Bell notions. The dates 1985–95 associated with quantum computing in figure 1.3 refer to the pre-philosophical association period of the field.

In quantum computing, alternative formulations of the basic probability amplitude for quantum entanglement are (see, for instance, Steane 1998)

$$|\psi\rangle = \frac{1}{\sqrt{2}}(|\uparrow\rangle|\downarrow\rangle - |\downarrow\rangle|\uparrow\rangle) \tag{1.10}$$

and

$$|\psi\rangle = \frac{1}{\sqrt{2}}(|1\rangle|0\rangle - |0\rangle|1\rangle) \tag{1.11}$$

or

$$|\psi\rangle = \frac{1}{\sqrt{2}}(|10\rangle - |01\rangle). \tag{1.12}$$

1.6 Fundamentals of Quantum Entanglement

Fundamentals of Quantum Entanglement is a focused monograph designed to deal almost directly with the origin and physics of the probability amplitude for quantum entanglement

$$|\psi\rangle = \frac{1}{\sqrt{2}}(|x\rangle_1|y\rangle_2 \pm |y\rangle_1|x\rangle_2) \tag{1.13}$$

in its minimal coverage, which is for two quanta ($n = 2$) and two propagation paths ($N = 2$), or $n = N = 2$. The approach taken to explain the physics is a generalized interferometric approach (Duarte 2013a, 2013b, 2014). The extended cases for $n = N = 2^1, 2^2, 2^4...2^r$ and $n = N = 3, 6$ are also reviewed.

The selected style of presentation is a series of dedicated brief chapters. The aim is to convey in a focused style the essence of the subject at hand so as to offer the reader the opportunity to contemplate the subject matter before moving to another topic. In the first part of the book the presentation is chronological.

Next, the titles of the chapters, beyond this introduction, are listed with a brief explanation of their content.

Chapter 2: *Dirac's contribution*: provides a very brief non-technical description of Dirac's pair theory (Dirac 1930) followed by an introduction to Dirac's *bra–ket* notation (Dirac 1939). Both of these contributions went on to impact the field of quantum entanglement.

Chapter 3: *The EPR paper*: reviews the contribution of Einstein *et al* (1935) that was the genesis of the philosophical path to quantum entanglement.

Chapter 4: *The Schrödinger papers*: reviews the Schrödinger papers that first mentioned the words *entanglement* and *disentanglement* in a quantum mechanical framework (Schrödinger 1935, 1936).

Chapter 5: *The Wheeler paper*: reviews the paper that provided the first lucid and transparent description, in words, of the iconic quantum entanglement experiment (Wheeler 1946).

Chapter 6: *The probability amplitude for quantum entanglement*: introduces the Pryce–Ward probability amplitude, $|\psi\rangle = (|x, y\rangle - |y, x\rangle)$, at the center of quantum entanglement physics and the experimental schematics of the original experiment (Pryce and Ward 1947, Ward 1949).

Chapter 7: *The quantum entanglement experiment*: provides a description of the concepts involved in the quantum entanglement experiment based on the probability amplitude $|\psi\rangle = (|x_1, y_2\rangle - |y_1, x_2\rangle)$.

Chapter 8: *The annihilation quantum entanglement experiments*: describes the first quantum entanglement experiments utilizing positron–electron annihilation, via $e^+e^- \rightarrow \gamma_1\gamma_2$, and detection utilizing gamma ray scattering configurations (Hanna 1948, Bleuler and Bradt 1948, Wu and Shaknov 1950).

Chapter 9: *The Bohm–Aharanov paper*: reviews the first paper that brought together the philosophical path, including aspects of local hidden variable theories, and the physics path. This paper became an additional point of origin, apart from EPR's work, for the mainstream literature on quantum entanglement (Bohm and Aharonov 1957).

Chapter 10: *Bell's theorem*: reviews the paper that provided a modern transparent proof that local hidden variable theories are incompatible with the predictions of quantum mechanics (Bell 1964).

Chapter 11: *Feynman's Hamiltonians*: outlines Feynman's contributions that led to equations of the form $|\psi\rangle = (|B\rangle - |A\rangle)$ but did not derive the probability amplitude for quantum entanglement (Feynman *et al* 1965).

Chapter 12: *The second Wu quantum entanglement experiment*: reviews a paper that recasts the quantum entanglement gamma ray scattering experiments in light of Bell's theorem (Kasday *et al* 1975).

Chapter 13: *The hidden variable theory experiments*: mentions efforts to extend Bell's theorem to make possible testing for local hidden variable theories in the optical domain (Clauser *et al.* 1969, Clauser and Horner 1974).

Chapter 14: *The optical quantum entanglement experiments*: describes the experiments of Aspect *et al* (1981, 1982a, 1982b) that were initiated in an effort to test for local hidden variable theories.

Chapter 15: *The quantum entanglement probability amplitude* 1947–91: reviews the various formats of $|\psi\rangle = (|x, y\rangle - |y, x\rangle)$ in the 1947–91 period.

Chapter 16: *The GHZ quantum entanglement probability amplitudes*: accounts for the reintroduction of $|\psi\rangle = (|x\rangle_1|y\rangle_2 - |y\rangle_1|x\rangle_2)$ into the mainstream of quantum entanglement research and the introduction of the GHZ probability amplitudes for four particles and three particles (Greenberger *et al* 1990).

Chapter 17: *The interferometric derivation of the quantum entanglement probability amplitude for* $n = N = 2$: describes the derivation of $|\psi\rangle = (|x\rangle_1|y\rangle_2 - |y\rangle_1|x\rangle_2)$ from a generalized N-slit interferometric perspective (Duarte 2013a, 2013b, 2014).

Chapter 18: *The interferometric derivation of the quantum entanglement probability amplitudes for* $n = N = 2^1, 2^2, 2^3...2^r$: describes the systematic derivation of probability amplitudes for $n = N = 2^1, 2^2, 2^3...2^r$ from the N-slit interferometric perspective (Duarte 2015, 2016, Duarte and Taylor 2017).

Chapter 19: *The interferometric derivation of the quantum entanglement probability amplitude for* $n = N = 3, 6$: describes the derivation of probability amplitudes for $n = N = 3, 6$ from the N-slit interferometric perspective (Duarte 2015, 2016).

Chapter 20: *What happens with the entanglement at* $n = 1$ *and* $N = 2$?: discusses reversibility from quantum entanglement probability amplitudes to interferometric probability amplitudes and vice versa. It also discusses the interferometric physics at $n = 1$ and $N = 2$.

Chapter 21: *The quantum entanglement probability amplitude and Bell's theorem*: describes how probabilities originating in the probability amplitude $|\psi\rangle = (|x\rangle_1|y\rangle_2 - |y\rangle_1|x\rangle_2)$ are used in conjunction with Bell's inequalities.

Chapter 22: *Cryptography via quantum entanglement*: describes how secure cryptographic communications, in free space, can be accomplished via the physics of quantum entanglement.

Chapter 23: *Quantum entanglement and teleportation*: explain the quantum entanglement concepts applied to achieve teleportation of quantum states.

Chapter 24: *Quantum entanglement and quantum computing*: provides a succinct overview on the use of quantum entanglement concepts, via $|\psi\rangle = (|x\rangle_1|y\rangle_2 - |y\rangle_1|x\rangle_2)$, to perform quantum logic operations. It also offers an additional avenue, via Pauli matrices, to derive $|\psi\rangle = (|x\rangle_1|y\rangle_2 - |y\rangle_1|x\rangle_2)$.

Chapter 25: *Space-to-space and space-to-Earth communications* via *quantum entanglement*: revisits the concept of quantum cryptography with an specific emphasis on satellite communications.

Chapter 26: *Quantum interferometric communications as an alternative to quantum entanglement?*: explains the physics of secure space-to-space communications using quantum interference principles.

Chapter 27: *Quanta sources for quantum entanglement*: reviews the various sources of quanta pairs exhibiting orthogonal polarizations, and provides an outlook on possible future sources.

Chapter 28: *More on quantum entanglement*: provides a pragmatic perspective of quantum entanglement divergent from philosophical concerns.

Chapter 29: *On the interpretation of quantum mechanics*: this is a discussion on the interpretation of quantum mechanics from a pragmatic perspective along the lines of thought of Dirac, Feynman, Lamb, van Kampen, and Ward (Dirac 1987, Feynman *et al* 1965, Lamb 1987, 2001, van Kampen 1987, Ward 2004).

Appendix A: Revisiting the EPR paper.

Appendix B: Revisiting the Pryce–Ward probability amplitude.

Appendix C: Classical and quantum interference.

Appendix D: Interferometers and their probability amplitudes.

Appendix E: Polarization rotators.

Appendix F: Vector products in quantum notation.

Appendix G: Trigonometric identities.

Appendix H: More on probability amplitudes.

Appendix I: From quantum principles to classical optics.

Appendix J: Introduction to Hamilton's quaternions.

1.7 Intent

Today, there are many young engineers working on many of the aspects, and applications, of quantum entanglement. Quantum mechanics works, the technology works, and this is a wonderful quantum world. However, many questions persist on the origin of quantum entanglement. Although various books and reviews expose the subject, they do so from the philosophical perspective. Moreover, they do so while introducing the probability amplitude $|\psi\rangle = (|x\rangle_1|y\rangle_2 - |y\rangle_1|x\rangle_2)$ out of the blue, without explaining its origin or physics. This is not the case with *Fundamentals of Quantum Entanglement*. Here, the origin of this wondrous probability amplitude is explained via the beautiful interferometric principles laid down by Dirac and championed by Feynman. In the minds of many readers questions will probably remain; however, it is hoped that the tools and ideas presented here will help those driven by curiosity to continue the path of discovery.

References

Aspect A, Grangier P and Roger G 1981 Experimental tests of realistic local theories via Bell's theorem *Phys. Rev. Lett.* **47** 460–3

Aspect A, Grangier P and Roger G 1982a Experimental realization of Einstein–Podolsky–Rosen–Bohm gedanken experiment: a new violation of Bell's inequalities *Phys. Rev. Lett.* **49** 91–4

Aspect A, Grangier P and Roger G 1982b Experimental test of Bell's inequality using time-varying analyzers *Phys. Rev. Lett.* **49** 1804–7

Bell J S 1964 On the Einstein–Podolsky–Rosen paradox *Physics* **1** 195–200

Bennett C H and Brassard G 1984 Quantum cryptography: public key distribution and coin tossing *Proc. of the IEEE Int. Conf. on Computer systems and Signal Processing* (Bangalore, India: IEEE)

Bennett C H, Brassard G, Crépeau C, Jozsa R, Peres A and Wootters W K 1993 Teleporting an unknown quantum state via dual classical and Einstein–Podolsky–Rosen channels *Phys. Rev. Lett.* **70** 1895–9

Bleuler E and Bradt H L 1948 Correlation between the states of polarization of the two quanta of annihilation radiation *Phys. Rev.* **73** 1398

Bohm D and Aharonov Y 1957 Discussion of experimental proof for the paradox of Einstein, Rosen, and Podolsky *Phys. Rev.* **108** 1070–6

Born M, Heisenberg W and Jordan P 1926 Zur quantenmechanik II *Z. Phys.* **35** 557–615

Born M and Jordan P 1925 Zur quantenmechanik *Z. Phys.* **34** 858–88

Clauser J F, Horner M A, Shimony A and Holt R A 1969 Proposed experiment to test local hidden variable theories *Phys. Rev. Lett.* **23** 880–3

Clauser J F and Horner M A 1974 Experimental consequences of objective local theories *Phys. Rev.* D **10** 526–35

Dalitz R H and Duarte F J 2000 John Clive Ward *Phys. Today* **53** 99–100

de Broglie L 1924 A tentative theory of light quanta *London Edinburgh Dublin Philos. Mag. J. Sci.* **47** 446–58

Dirac P A M 1930 On the annihilation of electrons and protons *Math. Proc. Camb. Philos. Soc.* **2** 361–75

Dirac P A M 1939 A new notation for quantum mechanics *Math. Proc. Camb. Philos. Soc.* **35** 416–8

Dirac P A M 1978 *The Principles of Quantum Mechanics* 4th edn (Oxford: Oxford University)

Dirac P A M 1987 The inadequacies of quantum field theory *Paul Adrien Maurice Dirac* ed B N Kursunoglu and E P Wigner (Cambridge: Cambridge University) ch 15

Duarte F J 2003 *Tunable Laser Optics* (New York: Elsevier)

Duarte F J 2012 The origin of quantum entanglement experiments based on polarization measurements *Eur. Phys. J. H* **37** 311–8

Duarte F J 2013a The probability amplitude for entangled polarizations: an interferometric approach *J. Mod. Opt.* **60** 1585–7

Duarte F J 2013b Tunable laser optics: applications to optics and quantum optics *Prog. Quantum Electron.* **37** 326–47

Duarte F J 2014 *Quantum Optics for Engineers* (New York: CRC)

Duarte F J 2015 *Tunable Laser Optics* 2nd edn (New York: CRC)

Duarte F J 2016 Secure space-to-space interferometric communications and its nexus to the physics of quantum entanglement *Appl. Phys. Rev.* **3** 041301

Duarte F J and Taylor T S 2017 Quantum entanglement probability amplitudes in multiple channels: an interferometric approach *Optik* **139** 222–30

Einstein A, Podolsky B and Rosen N 1935 Can quantum mechanical description of physical reality be considered complete? *Phys. Rev.* **47** 777–80

Ekert A K 1991 Quantum cryptography based on Bell's theorem *Phys. Rev. Lett.* **67** 661–3

Feynman R P 1985 Quantum mechanical computers *Opt. News* **11** 11–20

Feynman R P 1986 Quantum mechanical computers *Found. Phys.* **16** 507–31

Feynman R P and Hibbs A R 1965 *Quantum Mechanics and Path Integrals* (New York: McGraw-Hill)

Feynman R P, Leighton R B and Sands M 1965 *The Feynman Lectures on Physics* vol III (Reading: Addison-Wesley)

Ghirardi G C, Rimini A and Weber T 1987 Disentanglement of quantum functions: Answer to comment on 'Unified dynamics for microscopic and macroscopic systems' *Phys. Rev.* D **36** 3287–789

Greenberger D M, Horne M A, Shimony A and Zeilinger A 1990 Bell's theorem without inequalities *Am. J. Phys.* **58** 1131–43

Hanna R C 1948 Polarization of annihilation radiation *Nature* **162** 332

Heisenberg W 1925 Über quantenthoretische umdeutung kinematischer und mechanischer beziehungen *Z. Phys.* **33** 879–93

Kasday L R, Ullman J D and Wu C S 1975 Angular correlation of compton-scattered annihilation photons and hidden variables *Il Nuovo Cimento* **25** 633–61

Lamb W E 1987 *Schrödingers's cat Paul Adrien Maurice Dirac* ed B N Kursunoglu and E P Wigner (Cambridge: Cambridge University) ch 21

Lamb W E 1995 Anti-photon *Appl. Phys.* B **60** 77–84

Lamb W E 2001 Super classical quantum mechanics: the best interpretation of non-relativistic quantum mechanics *Am. J. Phys.* **69** 413–21

Moyal J E 1949 Quantum mechanics as a statistical theory *Proc. Camb. Philos. Soc.* **45** 99–124

Planck M 1901 Ueber das gesetz der energieverteilung im normalspectrum *Ann. Phys.* **309** 553–63

Pryce M L H and Ward J C 1947 Angular correlation effects with annihilation radiation *Nature* **160** 435

Schrödinger E 1926 An undulatory theory of the mechanics of atoms and molecules *Phys. Rev.* **28** 1049–70

Schrödinger E 1935 Discussion of probability relations between separated systems *Math. Proc. Camb. Philos. Soc.* **31** 555–63

Schrödinger E 1936 Probability relations between separated systems *Math. Proc. Camb. Philos. Soc.* **32** 446–52

Schumacher B 1995 Quantum coding *Phys. Rev. Lett.* **51** 2738–47

Snyder H S, Pasternack S and Hornbostel J 1948 Angular correlations of scattered annihilated radiation *Phys. Rev.* **73** 440–8

Steane A 1998 Quantum computing *Rep. Prog. Phys.* **61** 117–74

van Kampen N G 1987 Ten theorems about quantum mechanical measurements *Physica A* **153** 97–113

von Neumann J 1932 *Mathematische Grundlagen der Quanten-Mechanik* (Berlin: Springer)

Ward J C 1949 *Some Properties of the Elementary Particles* (Oxford: Oxford University)

Ward J C 2004 *Memoirs of a Theoretical Physicist* (Rochester: Optics Journal)

Wheeler J A 1946 Polyelectrons *Ann. N. Y. Acad. Sci.* **48** 219–38

Wilson A R, Lowe J and Butt D K 1976 Measurement of the relative planes of polarization of annihilation quanta as a function of separation distance *J. Phys. G: Nucl. Phys.* **2** 613–24

Wu C S and Shaknov I 1950 The angular correlation of scattered annihilation radiation *Phys. Rev.* **77** 136

Chapter 2

Dirac's contribution

The little-known contributions of Dirac to quantum entanglement are highlighted. The Dirac *bra–ket* notation, used throughout the monograph, is then introduced from an interferometric perspective together with the Dirac identities relevant to quantum entanglement.

2.1 Introduction

Paul Dirac made various titanic contributions to the development of quantum mechanics (Dirac 1926, 1927, 1928, 1929, 1930a, 1931). In addition to those well-known and celebrated contributions, he also made two other contributions that would eventually tremendously impact the physics of quantum entanglement; however, at the time of the disclosure of these contributions, the notion of and the words *quantum entanglement* were neither mentioned nor discussed. In other words, these were contributions silent about quantum entanglement but destined to have a significant future impact on this field.

In this chapter, a brief description of Dirac's pair theory is given while most of the chapter is devoted to introduce quantum interference *à la Dirac*. The reason for this will become apparent in the main body of the book but can be summarized in a brief sentence: interference is the essence of quantum entanglement.

More on Dirac can be found in Kragh's biography (Kragh 1990), Farmelo's biography (Farmelo 2009), and in a wonderful tribute book written by his contemporaneous colleagues (Kursunoglu and Wigner 1987).

2.2 Dirac's pair theory

In 1930 Dirac published a paper entitled 'On the annihilation of electrons and protons' (Dirac 1930b). The theory based on this paper helped to elucidate Anderson's discovery on the creation of pairs of electrons and positrons by electromagnetic radiation (Anderson 1932).

Dirac's paper discusses the emission of two quanta as the result of an annihilation process. This paper introduced the theoretical bases of what became known as *pair theory*.

For quantum entanglement, two quanta, propagating in opposite directions and each of which exhibiting orthogonal linear polarizations, are necessary. That these two quanta have orthogonal polarizations was first advanced by Wheeler (1946) and was theoretically confirmed by Pryce and Ward (1947) and Ward (1949).

2.3 Dirac's notation

Dirac introduced a notational alternative to quantum mechanics in a paper entitled 'A new notation for quantum mechanics' (Dirac 1939). In this notation, a *ket* vector is denoted by $|\rangle$, and a *bra* vector is denoted by $\langle|$. These vectors are mirror images of each other. Thus, the *probability amplitude* to go from the s plane to the x plane is described by the *bra–ket* $\langle x|s \rangle$, which is a *complex number*. It is important to note that the propagation from the s plane to the x plane is expressed *in reverse* by $\langle x|s \rangle$.

If the propagation of a single photon from plane s to plane x involves the passage through an intermediate plane j, then the probability amplitude becomes (Feynman *et al* 1965)

$$\langle x|s \rangle = \langle x| j \rangle \langle j|s \rangle. \tag{2.1}$$

If there are N alternatives for j as in the case of a transmission diffraction grating, that is $j = 1, 2, 3 \ldots N$, then an N-slit interferometer is configured (as illustrated in figure 2.1) and the corresponding probability amplitude becomes

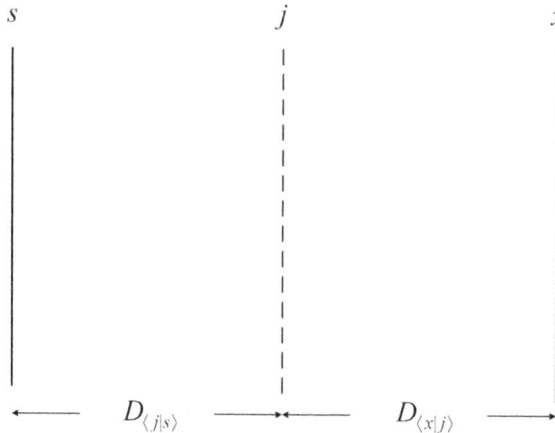

Figure 2.1. Schematic configuration to describe single-photon propagation from plane s to x via the transmission interferometric plane j. Plane j is configured by an array on N-slits. Interference takes place between plane j and the interferometric, or detection, plane x. The intra–interferometric distance from plane j to x is denoted as $D_{\langle x|j \rangle}$. This schematic is also applicable to illumination by a population of indistinguishable photons.

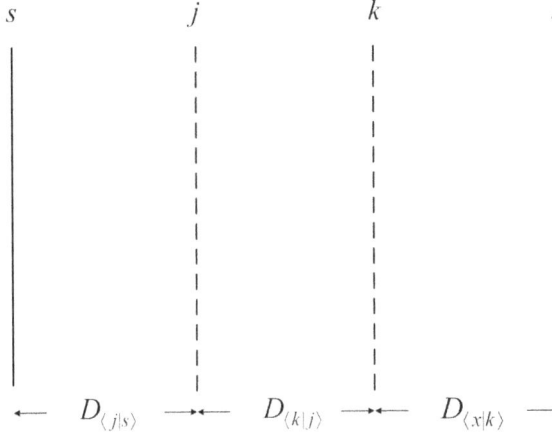

Figure 2.2. Schematic configuration to describe single-photon propagation from plane s to x via the transmission interferometric planes j and k. Planes j and k are configured by N-slits arrays. *Measurable interference* takes place between plane k and the interferometric, or detection, plane x. The intra–interferometric distances are denoted by $D_{\langle k|j\rangle}$ and $D_{\langle x|k\rangle}$. This schematic is also applicable to illumination by a population of indistinguishable photons.

$$\langle x|s\rangle = \sum_{j=1}^{N}\langle x|j\rangle\langle j|s\rangle. \tag{2.2}$$

At this stage, it should be indicated that equation (2.2) represents far more than a notational alternative. It is a highly utilitarian and profound statement in statistical physics, as will become apparent in later chapters of this book. The distance from the plane s to j is denoted as $D_{\langle j|s\rangle}$ while the interferometric distance from plane j to x is denoted as $D_{\langle x|j\rangle}$. If a further intermediate plane k, in addition to j, is involved, as illustrated in figure 2.2, then the probability amplitude becomes

$$\langle x|s\rangle = \sum_{k=1}^{N}\sum_{j=1}^{N}\langle x|k\rangle\langle k|j\rangle\langle j|s\rangle. \tag{2.3}$$

Here, the intra-interferometric distances are denoted by $D_{\langle k|j\rangle}$ and $D_{\langle x|k\rangle}$. If a third intermediate plane l, of N-slits, is included as illustrated in figure 2.3, then the probability amplitude becomes

$$\langle x|s\rangle = \sum_{l=1}^{N}\sum_{k=1}^{N}\sum_{j=1}^{N}\langle x|l\rangle\langle l|k\rangle\langle k|j\rangle\langle j|s\rangle. \tag{2.4}$$

Here, the intra-interferometric distances are denoted by $D_{\langle k|j\rangle}$, $D_{\langle l|k\rangle}$, and $D_{\langle x|l\rangle}$.

2.4 Dirac's notation in N-slit interferometers

In his book, Dirac refers directly to practical, workable, macroscopic interferometry using quantum principles. He refers to a monochromatic beam composed of a large

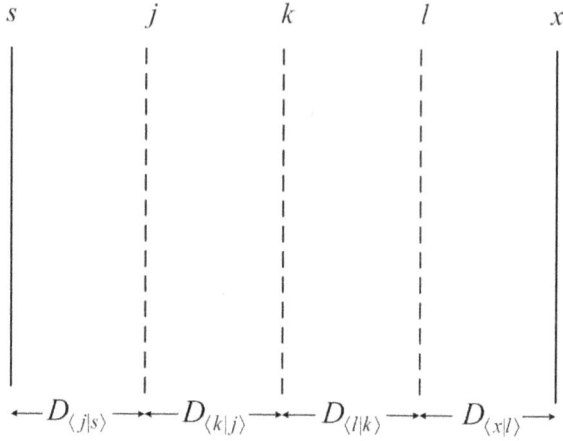

Figure 2.3. Schematic configuration to describe single-photon propagation from plane s to x via the transmission interferometric planes j, k, and l. Planes j, k, and l are configured by N-slits arrays. *Measurable interference* takes place between plane l and the interferometric, or detection, plane x. The intra–interferometric distances are denoted by $D_{\langle k|j\rangle}$, $D_{\langle l|k\rangle}$, and $D_{\langle x|l\rangle}$. This schematic is also applicable to illumination by a population of indistinguishable photons.

number of photons. A single photon in this beam is distributed into its probability amplitude components and these components are made to interfere. It is in this discussion that he made his famous statement 'each photon then interferes only with itself' (Dirac 1978). His description is perfectly applicable to a population of indistinguishable photons as would be the case of interference using a single-transverse-mode single-longitudinal-mode narrow-linewidth laser (Duarte 2003).

In order to transition from a purely mathematical probability amplitude to a *probability*, $\langle x|s\rangle$ must be multiplied with its complex conjugate $\langle x|s\rangle^*$ according to Born's rule (Born 1926), that is,

$$\langle x|s\rangle\langle x|s\rangle^* = \left(\sum_{j=1}^{N}\langle x|j\rangle\langle j|s\rangle\right)\left(\sum_{j=1}^{N}\langle x|j\rangle\langle j|s\rangle\right)^*. \qquad (2.5)$$

Following Dirac, the individual probability amplitudes can be represented by *wave functions of ordinary wave optics* (Dirac 1978). Hence, the individual probability amplitudes can be written as

$$\langle j|s\rangle = \Psi(r_{j,s})e^{-i\theta_j} \qquad (2.6)$$

$$\langle x|j\rangle = \Psi(r_{x,j})e^{-i\phi_j} \qquad (2.7)$$

where θ_j and ϕ_j are the phase terms associated with the incidence and diffraction waves, respectively. Using equations (2.6) and (2.7), the propagation probability amplitude becomes

$$\langle x|s \rangle = \sum_{j=1}^{N} \Psi(r_j) e^{-i\Omega_j} \tag{2.8}$$

where

$$\Psi(r_j) = \Psi(r_{x,j}) \Psi(r_{j,s}) \tag{2.9}$$

and the phase is given by

$$\Omega_j = (\theta_j + \phi_j). \tag{2.10}$$

Using equation (2.8) to perform the complex conjugate product

$$|\langle x|s \rangle|^2 = \langle x|s \rangle \langle x|s \rangle^* \tag{2.11}$$

leads to the generalized interferometric equation in complex form:

$$|\langle x \mid s \rangle|^2 = \sum_{j=1}^{N} \Psi(r_j) \sum_{m=1}^{N} \Psi(r_m) e^{i(\Omega_m - \Omega_j)}. \tag{2.12}$$

Expanding equation (2.12) and using the identity

$$2\cos(\Omega_m - \Omega_j) = e^{-i(\Omega_m - \Omega_j)} + e^{i(\Omega_m - \Omega_j)} \tag{2.13}$$

leads to the explicit form of the generalized interferometric equation in one dimension (Duarte and Paine 1989, Duarte 1991, 1993):

$$|\langle x|s \rangle|^2 = \sum_{j=1}^{N} \Psi(r_j)^2 + 2\sum_{j=1}^{N} \Psi(r_j) \left(\sum_{m=j+1}^{N} \Psi(r_m)\cos(\Omega_m - \Omega_j) \right). \tag{2.14}$$

The reader should notice that the interferometric equations (2.5), (2.12), and (2.14) are completely equivalent. These interferometric equations are *probabilities*. Equation (2.5) is purely applicable to single-photon propagation whilst equations (2.12) and (2.14) are also applicable to the propagation of *populations of indistinguishable photons*. Populations of indistinguishable photons are available from highly coherent sources such as single-transverse-mode single-longitudinal-mode narrow-linewidth lasers. The reader should also be aware that despite this interferometric trinity, equation (2.14) is sometimes confused with equations of intensity used in classical interference optics. This topic is discussed further in appendix C.

On illumination: 'all the indistinguishable photons illuminate the array of N-slits, or grating, simultaneously. If only one photon propagates, at any given time, then that individual photon illuminates the whole array of N-slits simultaneously' (Duarte 2003).

The application of the Dirac notation to practical interferometric systems is described by Duarte (1993). Explicit and extensive discussions in a review style are given by Duarte (2014, 2015).

The generalized interferometric equation in its complex form, as given in equation (2.12), can be expressed in two dimensions as (Duarte 1995, 2014)

$$|\langle x|s\rangle|^2 = \sum_{z=1}^{N}\sum_{y=1}^{N}\Psi(r_{zy})\sum_{q=1}^{N}\sum_{p=1}^{N}\Psi(r_{pq})e^{i(\Omega_{qp}-\Omega_{zy})}, \tag{2.15}$$

and in three dimensions as (Duarte 1995, 2014)

$$|\langle x|s\rangle|^2 = \sum_{z=1}^{N}\sum_{y=1}^{N}\sum_{x=1}^{N}\Psi(r_{zyx})\sum_{q=1}^{N}\sum_{p=1}^{N}\sum_{r=1}^{N}\Psi(r_{qpr})e^{i(\Omega_{qpr}-\Omega_{zyx})}. \tag{2.16}$$

Note: the book by Dirac entitled *The Principles of Quantum Mechanics*, which is extensively used as a reference in this book, was first published in 1930. The *bra–ket* notation was incorporated in its third edition (1947). Here, the eighth printing of the fourth edition (1978) is being used. It is interesting to note that in his introductory paper on the subject, Dirac does not explain the origin of his *bra–ket* notation. In the year 2000 I asked Richard Dalitz, a particle physicist who took lectures from Dirac at Cambridge and interacted with him for years, if Dirac ever explained how he came about his quantum notation. His answer was ... 'no' (Dalitz 2000).

2.5 Semi coherent interference

Equations (2.5), (2.12), and (2.14) apply to single-photon interference and interference using a population of indistinguishable photons, which is equivalent to interference whilst utilizing a narrow-linewidth laser. As discussed by Duarte (2014, 2015), semi coherent interference can be mathematically represented by

$$\sum_{\lambda=\lambda_1}^{\lambda_n}|\langle x|s\rangle|_\lambda^2 = \sum_{\lambda=\lambda_1}^{\lambda_n}\left(\sum_{j=1}^{N}\Psi(r_j)_\lambda\sum_{m=1}^{N}\Psi(r_m)_\lambda e^{i(\Omega_m-\Omega_j)}\right). \tag{2.17}$$

Physically, the meaning of a semi coherent measurement is that the detector registers an interferometric pattern for a given quanta of a defined wavelength, λ_1 let us say, and then a different pattern for λ_2, and so on. The overall measurement is a cumulative interferometric pattern corresponding to λ_1, λ_2, λ_3... λ_n (see chapter 28). This explains why the interferograms derived from semi coherent illumination are broader and less defined than the interferograms recorded using highly coherent radiation such as narrow-linewidth laser illumination (Duarte 2015).

2.6 From quantum probabilities to measurable intensities

It should be re-emphasized that the generalized interferometric probability equations, as given in equations (2.5), (2.12), and (2.14)–(2.16), *are probabilities*. As explained in the literature (Duarte 2004), these equations correspond to spatial probability distributions that determine the spatial boundaries for the arrival of quanta at the interference plane or detection surface. The arrival of an individual photon, with an energy $E = h\upsilon$, is registered at the detector by the creation of a charge within the spatial boundaries of the probability distribution $\langle x|s\rangle\langle x|s\rangle^*$. For the case of populations of indistinguishable photons, a cumulative charge distribution within the spatial boundaries of $\langle x|s\rangle\langle x|s\rangle^*$ is registered at the detector (Duarte 2004). This cumulative charge distribution is what the experimenter observes as the spatial pattern of the photon count or the spatial distribution of the intensity.

Furthermore, following Feynman et al (1965) and Lamb and colleagues (Sargent et al 1974), it can be shown that the spatial intensity distribution, at a given frequency ν, is proportional to the spatial probability distribution $\langle x|s\rangle\langle x|s\rangle^*$ (Duarte 2014):

$$\mathcal{I}(\nu) \propto |\langle x|s\rangle\langle x|s\rangle^*| \tag{2.18}$$

or

$$\mathcal{I}(\nu) = K|\langle x|s\rangle\langle x|s\rangle^*| \tag{2.19}$$

where the dimensionless $\langle x|s\rangle\langle x|s\rangle^*$ provides all the spatial information. Here, the intensity $\mathcal{I}(\nu)$ has the units of J s^{-1} m^{-2} or W m^{-2}.

2.7 Dirac's identities

In this section a few of Dirac's notational identities, useful to the subject at hand, are listed. Notation was an extraordinarily important topic in Dirac's physics and he was meticulous and highly precise in the equations he wrote. Here, these identities are collected from various sections of Dirac's book and are compiled roughly in the same order as presented by Duarte (2014), where a more complete set of identities is presented. In the style of Dirac, no attempt is made to explain any of this:

$$\langle \phi|\psi\rangle = \langle \psi|\phi\rangle^* \tag{2.20}$$

$$\langle \phi|\psi\rangle = \langle \phi|j\rangle\langle j|\psi\rangle. \tag{2.21}$$

Some abstraction identities:

$$|\psi\rangle = |j\rangle\langle j\,|\psi\rangle \tag{2.22}$$

$$\langle \chi|A|\phi\rangle = \langle \chi|i\rangle\langle i|A|j\rangle\langle j|\phi\rangle \tag{2.23}$$

$$A = |i\rangle\langle i|\,A|j\rangle\langle j| \tag{2.24}$$

$$A|\phi\rangle = |i\rangle\langle i|\,A|j\rangle\langle j|\,\phi\rangle. \tag{2.25}$$

If

$$\langle i|\phi\rangle = C_i \tag{2.26}$$

$$\langle i|\chi\rangle = D_i \tag{2.27}$$

then

$$|\phi\rangle = \sum_i |i\rangle C_i \qquad (2.28)$$

$$|\chi\rangle = \sum_i |i\rangle D_i \qquad (2.29)$$

and

$$\langle\chi| = \sum_i D_i^*\langle i|, \qquad (2.30)$$

which is the abstracted expression of probability amplitude

$$\langle\chi|\phi\rangle = \sum_{ij} D_j^*\langle j|i\rangle C_i. \qquad (2.31)$$

Using $\langle j|i\rangle = \delta_{ij}$,

$$\langle\chi|\phi\rangle = \sum_i D_i^* C_i. \qquad (2.32)$$

As suggested by Dirac (Feynman *et al* 1965), equation (2.32) can be written as

$$| = \sum_i |i\rangle\langle i|. \qquad (2.33)$$

To the preceding compendium the following identities are also relevant (Dirac 1978):

$$|a\rangle|b\rangle = |b\rangle|a\rangle = |ab\rangle \qquad (2.34)$$

$$|a\rangle|b\rangle|c\rangle\dots = |abc\dots\rangle \qquad (2.35)$$

and the property of being commutative is (see appendix H)

$$|a\rangle|b\rangle = |b\rangle|a\rangle = |ab\rangle = |ba\rangle. \qquad (2.36)$$

Relevant to the quantum entanglement of several similar particles are the following identities (Dirac 1978):

$$|a_1\rangle|b_2\rangle|c_3\rangle\dots|g_n\rangle = |a_1 b_2 c_3 \dots g_n\rangle \qquad (2.37)$$

$$|b_1\rangle|a_2\rangle|c_3\rangle\dots|g_n\rangle = |b_1 a_2 c_3 \dots g_n\rangle. \qquad (2.38)$$

Identity (2.37) can also be expressed as

$$|a\rangle_1|b\rangle_2|c\rangle_3\dots|g\rangle_n = |a_1 b_2 c_3 \dots g_n\rangle \qquad (2.39)$$

and by introducing

$$|X\rangle = |a_1 b_2 c_3 \dots g_n\rangle \qquad (2.40)$$

identity (2.39) can be rewritten as

$$|X\rangle = |a\rangle_1|b\rangle_2|c\rangle_3\dots |g\rangle_n. \tag{2.41}$$

The identity expressed in equation (2.41) is crucial to the derivation of the quantum entanglement probability amplitude from interferometric premises, as will be shown in chapters 17 and 18. Further aspects of quantum notation are examined in appendix H.

References

Anderson C D 1932 Energies of cosmic-ray particles *Phys. Rev.* **42** 405–21

Born M 1926 Zur quantenmechanik der stoßvorgänge *Z. Phys.* **37** 863–7

Dalitz R H 2000 Private communication

Dirac P A M 1926 On the theory of quantum mechanics *Proc. R. Soc.* A **112** 661–77

Dirac P A M 1927 The quantum theory of the emission and absorption of light *Proc. R. Soc.* A **114** 243–65

Dirac P A M 1928 The quantum theory of the electron *Proc. R. Soc.* A **117** 610–24

Dirac P A M 1929 Quantum mechanics of many-electron systems *Proc. R. Soc.* A **123** 714–33

Dirac P A M 1930a A theory of electrons and protons *Proc. R. Soc.* A **126** 360–5

Dirac P A M 1930b On the annihilation of electrons and protons *Math. Proc. Camb. Philos. Soc.* **2** 361–75

Dirac P A M 1931 Quantised singularities in the electromagnetic filed *Proc. R. Soc.* A **133** 60–72

Dirac P A M 1939 A new notation for quantum mechanics *Math. Proc. Camb. Philos. Soc.* **35** 416–8

Dirac P A M 1978 *The Principles of Quantum Mechanics* 4th edn (Oxford: Oxford University)

Duarte F J 1991 Dispersive dye lasers *High Power Dye Lasers* ed F J Duarte (Berlin: Springer) ch 2

Duarte F J 1993 On a generalized interference equation and interferometric measurements *Opt. Commun.* **103** 8–14

Duarte F J 1995 Interferometric imaging *Tunable Laser Applications* 1st edn ed F J Duarte (New York: Marcel Dekker) ch 5

Duarte F J 2003 *Tunable Laser Optics* (New York: Elsevier)

Duarte F J 2004 Comment on 'Reflection, refraction and multislit interference' *Eur. J. Phys.* **25** L57–8

Duarte F J 2014 *Quantum Optics for Engineers* (New York: CRC)

Duarte F J 2015 *Tunable Laser Optics* 2nd edn (New York: CRC)

Duarte F J and Paine D J 1989 Quantum mechanical description of N-slit interference phenomena *Proc. of the Int. Conf. on Lasers'88* ed R C Sze and F J Duarte (McLean, VA: STS Press), pp 42–7

Farmelo G 2009 *The Strangest Man: The Hidden Life of Paul Dirac, Mystic of the Atom* (New York: Basic Books)

Feynman R P, Leighton R B and Sands M 1965 *The Feynman Lectures on Physics* vol III (Reading, MA: Addison-Wesley)

Kragh H S 1990 *Dirac: A Scientific Biography* (Cambridge: Cambridge University Press)

Kursunoglu B N and Wigner E P (ed) 1987 *Paul Adrien Maurice Dirac* (Cambridge: Cambridge University Press)

Pryce M L H and Ward J C 1947 Angular correlation effects with annihilation radiation *Nature* **160** 435

Sargent M, Scully M O and Lamb W E 1974 *Laser Physics* (Reading, MA: Addison-Wesley)

Ward J C 1949 *Some Properties of the Elementary Particles* (Oxford: Oxford University)

Wheeler J A 1946 Polyelectrons *Ann. N. Y. Acad. Sci.* **48** 219–38

Chapter 3

The Einstein–Podolsky–Rosen (EPR) paper

The famous Einstein–Podolsky–Rosen (EPR) paper, suggesting that quantum mechanics is an incomplete theory, is briefly discussed in the context of quantum entanglement. Also, Einstein's definition of a correct theory is quoted.

3.1 Introduction

The EPR paper entitled 'Can quantum mechanical description of physical reality be considered complete?' (Einstein *et al* 1935) is one of the most cited papers in physics. In it, Einstein and colleagues question the suitability of the wave function, which is synonymous to the probability amplitude, to provide a complete description of physical reality. Therefore, in this chapter the EPR paper is revisited and its main questions and teachings are restated in an effort to transparently highlight its significance.

It should be noted that in the EPR paper no mention is made of the words 'quantum entanglement'. The genius of the EPR paper is that it considered two quantum systems that interact in an initial $0 \leqslant t \leqslant T$ time frame and cease to interact at times $t > T$. As will be explained in the following chapters, the association of EPR with the polarization entanglement of two quanta propagating in opposite directions would come later via Bohm and Aharanov (1957).

3.2 EPR's doubts on quantum mechanics

By far, the main concern of Einstein as expressed in the EPR paper was that the wave function ψ did not provide a complete description of physical reality. Since ψ has the same physical meaning as the probability amplitude, it is reasonable to assume that Einstein's doubts also extended to the probability amplitude.

The EPR paper is divided into two segments. In the first segment, EPR conclude that *'when the momentum of the particle is known, its coordinate has no physical reality ...'* (see figure 3.1), and next they pose the alternative, that either '(1) *the quantum mechanical description of reality given by the wave function is not complete*'

doi:10.1088/2053-2563/ab2b33ch3

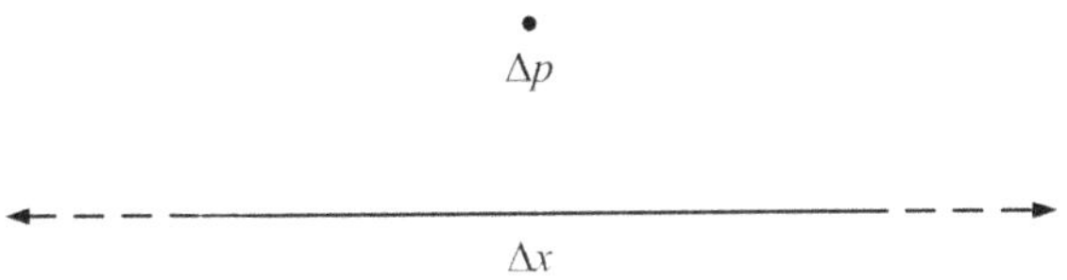

Figure 3.1. 'When *the momentum of the particle is known, its coordinate has no physical reality …*' (Einstein *et al* 1935). In other words, when the uncertainty in the measurement of p is infinitesimally small, that is $\Delta p \approx 0$, the uncertainty in x, that is Δx, is unmeasurably large.

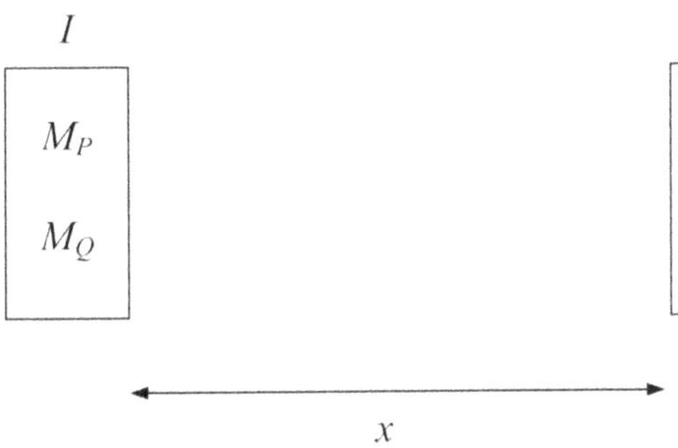

Figure 3.2. The quantities P and Q in system II depend on measurements M_P and M_Q performed in system I: 'No reasonable description of reality could be expected to permit this' (Einstein *et al* 1935).

or '(2) *when the operators corresponding to two physical quantities do not commute the two quantities cannot have simultaneous reality*' (Einstein *et al* 1935).

The second section of the EPR paper, which is directly relevant to quantum entanglement situations, can be succinctly summarized via their argument that as a result of two sequential measurements performed on a first system, a second system is left in two different states. They then emphasize that since the systems were not interacting with each other, no 'real change' can be induced over the second system by the measurements conducted on the first system (Einstein *et al* 1935).

Furthermore, EPR consider two systems while assigning the quantities P and Q to the second system. Here, their conclusion is that P and Q depend on the measurements performed in the first system, which is decoupled from the second system; see figure 3.2. Hence, they conclude: 'No reasonable description of reality could be expected to permit this' (Einstein *et al* 1935).

In summary, Einstein *et al* (1935) articulate a clear and explicit skepticism on the predictions of quantum mechanics. The conclusions reached by Einstein and his colleagues are considered further, from a different perspective, in appendix A. That discussion is provided from the perspective of the uncertainty principle (Duarte 2014), as defined by Dirac (1978) and Feynman *et al* (1965).

3.3 EPR's definition of a correct theory

Beyond the discussion of the effect of one quantum system on a second separated, and even distant, quantum system, EPR provides a lucid and landmark definition of what a theory should be: 'The correctness of the theory is judged by the degree of agreement between the conclusions of the theory and human experience. This experience … in physics takes the form of experiment and measurement' (Einstein *et al* 1935). In this book, this definition will be referred to as *EPR's definition of a correct theory*.

References

Bohm D and Aharanov Y 1957 Discussion of experimental proof for the paradox of Einstein, Rosen, and Podolsky *Phys. Rev.* **108** 1070–6

Dirac P A M 1978 *The Principles of Quantum Mechanics* 4th edn (Oxford: Oxford University)

Duarte F J 2014 *Quantum Optics for Engineers* (New York: CRC)

Einstein A, Podolsky B and Rosen N 1935 Can quantum mechanical description of physical reality be considered complete? *Phys. Rev.* **47** 777–80

Feynman R P, Leighton R B and Sands M 1965 *The Feynman Lectures on Physics* vol III (Reading, MA: Addison-Wesley)

Chapter 4

The Schrödinger papers

The Schrödinger papers that introduced the words *entanglement* and *disentanglement* in the context of the EPR paper are discussed.

4.1 Introduction

In 1935 and 1936 Schrödinger wrote two sequential papers on separated systems in reference to the EPR paper (Einstein *et al* 1935). It is in his first paper that Schrödinger introduced the concept of entanglement and the word itself: *entanglement* (Schrödinger 1935). The second paper is a follow up to the first paper (Schrödinger 1936). Neither paper refers to entangled polarization states of counter-propagating quanta.

4.2 The first Schrödinger paper

The first Schrödinger paper is entitled 'Discussion of probability relations between separated systems' (Schrödinger 1935). In it, Schrödinger provides a philosophical, and also physical, discussion of two separated systems that are entangled, and does reference the EPR paper. In this regard he uses the word 'disconcerting' to describe the fact that there is no independence between systems that were initially entangled and that are no longer in physical proximity: 'after re-establishing one ψ-function … the other one can be inferred simultaneously' (Schrödinger 1935). Lack of independence is associated with the fact that measurements in one system determine the wave function in the other system. And then Schrödinger adds that, 'It is rather discomforting that the theory should allow a system to be steered or piloted into one or the other type of state at the experimenter's mercy in spite of his having no access to it' (see figure 4.1).

This 1935 Schrödinger paper makes no mention of entangled polarizations of two counter-propagating quanta.

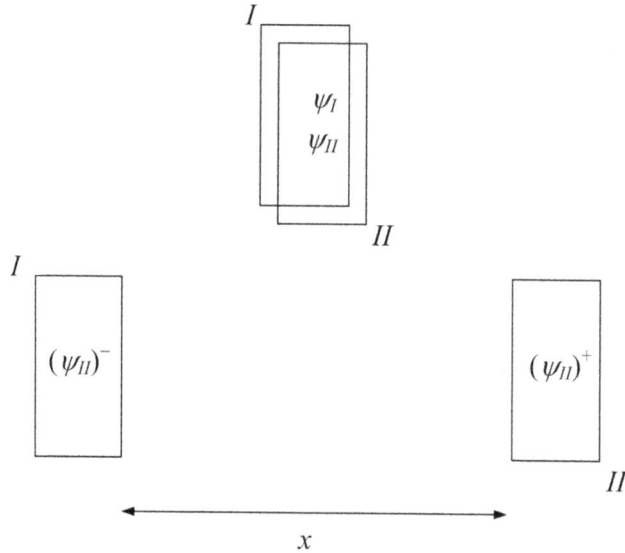

Figure 4.1. (a) According to Schrödinger, at first, systems I and II are entangled. (b) Following disentanglement, once the wave function in system II, $(\psi_{II})^+$, is determined, the wave function in system I, $(\psi_{II})^-$, can be 'inferred simultaneously'; or vice versa.

4.3 The second Schrödinger paper

The second Schrödinger paper is entitled 'Probability relations between separated systems' (Schrödinger 1936). In it, Schrödinger extends and reaffirms the nature of the discussion initiated in his first paper and points to the current interpretation of quantum mechanics that 'obliges' physicists to accept that, in a situation involving a system with two separated parts, performance of an experimenter's measurement on one part (part I) determines the state or wave function of the other part (part II) of the system. He further highlights the fact that the effect described above occurs in the absence of a direct interaction ('non-interference' as he labels it) with part II by the experimenter performing the measurement on part I. Specifically, the effect alluded by Schrödinger is that the state or wave function at part II of the system depends directly on the measurement performed by the experimenter on part I. Thus, he states that the experimenter conducting measurements at part I 'controls [the] future state' of part II of the system without 'touching' it (Schrödinger 1936). He continues to add: 'In this paper it will be shown that the control, with the indirect measurement, is in general not only *as* complete but even more complete'.

Note that the term *interference* has a definite and crucial meaning in quantum physics (Feynman *et al* 1965). In the reasoning by Schrödinger, the word 'non-interference' appears to be utilized as synonymous to 'non-interacting' and in the absence of physical touching and not in the context of quantum interference as defined by Dirac (1978) and Feynman *et al* (1965).

Furthermore, this 1936 Schrödinger paper makes no mention of entangled polarizations of two counter-propagating quanta.

References

Dirac P A M 1978 *The Principles of Quantum Mechanics* 4th edn (Oxford: Oxford University)

Einstein A, Podolsky B and Rosen N 1935 Can quantum mechanical description of physical reality be considered complete? *Phys. Rev.* **47** 777–80

Feynman R P, Leighton R B and Sands M 1965 *The Feynman Lectures on Physics* vol III (Reading: Addison-Wesley)

Schrödinger E 1935 Discussion of probability relations between separated systems *Math. Proc. Camb. Philos. Soc.* **31** 555–63

Schrödinger E 1936 Probability relations between separated systems *Math. Proc. Camb. Philos. Soc.* **32** 446–52

Chapter 5

Wheeler's paper

The 1946 paper by Wheeler, based on Dirac's pair theory, provided the first clear and transparent definition of the quantum entanglement of two quanta propagating in opposite directions and is introduced here.

5.1 Introduction

John Archival Wheeler was familiar with Dirac's pair theory (Dirac 1930) and he was the first to elucidate in clear and transparent English its significance to what would eventually become the field of quantum entanglement although he did not use the word entanglement in his paper. He was also the first to suggest, in prose, an experimental configuration to test the pair theory (Wheeler 1946). In this regard, Wheeler's paper was essential to initiate the purely physical developments of quantum entanglement, devoid of philosophical arguments, that would lead to the discovery of the probability amplitude for quantum entanglement (Pryce and Ward 1947, Ward 1949).

5.2 Wheeler's paper's significance to quantum theory

Wheeler's paper was entitled 'Polyelectrons' (Wheeler 1946). In his paper on electron–positron annihilation, Wheeler explained that the dominant mechanism of annihilation was $e^+e^- \to \gamma\gamma$ involving singlet states. A further important condition is that the pairs, prior to annihilation, should have zero relative angular momentum. Wheeler added that there is an analogous polarization phenomenon in the $\gamma\gamma$ pair created by the e^+e^- annihilation. Following this preamble, Wheeler added the crucial and transparent paragraph that describes the pure essence of quantum entanglement: 'According to the pair theory, *if one of these photons is linearly polarized in one plane, then the photon that goes off in the opposite direction with equal momentum is linearly polarized in the perpendicular plane*' (Wheeler 1946) … our italics. This concept is illustrated in figure 5.1.

doi:10.1088/2053-2563/ab2b33ch5

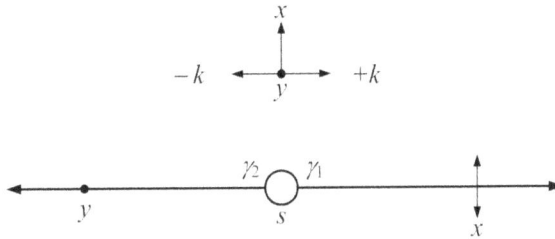

Figure 5.1. Diagrammatic rendition of Wheeler's description of the $e^+e^- \rightarrow \gamma_1\gamma_2$ experiment. Here, γ_1 and γ_2 are emitted in the $+k$ and $-k$ directions. If γ_1 is polarized in the $|x\rangle$ state, then γ_2 is polarized in the $|y\rangle$ state. In other words, the emitted quanta exhibit polarizations orthogonal to each other.

Here it should be noted that the explicit theory behind Wheeler's statement had not been disclosed. Also, it was that crucial statement that most likely guided Pryce and Ward to derive the probability amplitude for quantum entanglement (Pryce and Ward 1947, Ward 1949).

It should also be mentioned that Wheeler's paper did not refer to the philosophical arguments being developed in parallel, on quantum entanglement, by Einstein and Schrödinger (Einstein *et al* 1935, Schrödinger 1935).

5.3 Wheeler's paper's significance to quantum experiments

Furthermore, the Wheeler paper provided a nonschematic guide to a proposed annihilation–scattering experiment to test pair theory. In it, he describes a source that generates simultaneous quanta, with energy mc^2, in opposite directions via the annihilation of positrons. At the end of each emission channel a selective photon scatter is deployed prior to a photon counter. The angular position of the counters can be varied at will and coincidences are recorded when the angular orientation of the counters are identical and orthogonal to each other (Wheeler 1946).

The description above served as an explicit guide to Pryce and Ward (1947) to provide the first experimental diagram to compare the theory of quantum entanglement to experimental data (Pryce and Ward 1947). All subsequent optical experiments followed the same configurational strategy except that the scatters are replaced by polarizer analyzers, which are rotated instead of the photon counters.

References

Dirac P A M 1930 On the annihilation of electrons and protons *Math. Proc. Camb. Philos. Soc.* **2** 361–75

Einstein A, Podolsky B and Rosen N 1935 Can quantum mechanical description of physical reality be considered complete? *Phys. Rev.* **47** 777–80

Pryce M L H and Ward J C 1947 Angular correlation effects with annihilation radiation *Nature* **160** 435

Schrödinger E 1935 Discussion of probability relations between separated systems *Math. Proc. Camb. Philos. Soc.* **31** 555–63

Ward J C 1949 *Some Properties of the Elementary Particles* (Oxford: Oxford University)

Wheeler J A 1946 Polyelectrons *Ann. N. Y. Acad. Sci.* **48** 219–38

Chapter 6

The probability amplitude for quantum entanglement

The 1947 work of Pryce and Ward that first utilized the probability amplitude for quantum entanglement $|\psi\rangle = (|x_1, y_2\rangle - |y_1, x_2\rangle)$ in the context of an $e^+e^- \to \gamma_1\gamma_2$ experiment is introduced and discussed. Ward's derivation of this iconic equation is described.

6.1 Introduction

This chapter focuses on the works of Pryce and Ward (1947) and Ward (1949). The work of these physicists was inspired by the paper of Wheeler (1946). A relevant historical note here is that John Clive Ward was the doctoral student of Maurice Henry Lecorney Pryce, who was a student of Ralph Howard Fowler. R H Fowler was also the doctoral supervisor of Paul Adrien Maurice Dirac. This note hints at the quantum mechanics knowledge path that greatly benefited a young J C Ward born in 1924.

6.2 The Pryce–Ward paper

The Pryce–Ward paper, inspired by the previous work of Wheeler (1946), is a remarkable and succinct publication about a half-page long, on the emission by positron–electron annihilation yielding two quanta (γ_1 and γ_2), with orthogonal polarizations, emitted in opposite directions ($+k$ and $-k$). This process can be best illustrated by (Dalitz and Duarte 2000)

$$e^+e^- \to \gamma_1\gamma_2. \tag{6.1}$$

The Pryce and Ward paper came from the Dirac–Wheeler route in the complete absence of philosophical concerns and without using the word entanglement. Here, the theoretical and experimental significance of this paper are examined.

doi:10.1088/2053-2563/ab2b33ch6

The Pryce–Ward contribution is analyzed from a historical perspective in appendix B.

6.2.1 Theoretical legacy of the Pryce–Ward paper

The Pryce and Ward paper is all about providing a theoretical expression to predict the angular correlation of the polarizations of quanta γ_1 and γ_2 propagating in the opposite directions $+k$ and $-k$, as illustrated in figure 6.1. For completeness, it should be indicated that this condition of counter propagation, on the same optical axis, fulfills the condition of having opposite momenta via $|p| = \hbar k$, where k is the wave number.

Since this was not an experiment in the visible–near-infrared region of the electromagnetic spectrum, scatterers rather than polarizers were used, thus making the experimental–theoretical interaction far more complex than the optical experiments of this class that followed in the 1980s and onward. In reference to figure 6.1, Pryce and Ward derived the differential double cross-section as

$$\frac{r_0^4 d\Omega_1 d\Omega_2}{16}$$
$$\times \left[\frac{((1 - \cos\theta_1)^3 + 2)((1 - \cos\theta_2)^3 + 2)}{(2 - \cos\theta_1)^3(2 - \cos\theta_2)^3} - \frac{(\sin^2\theta_1 \sin^2\theta_2)\cos 2(\varphi_1 - \varphi_2)}{(2 - \cos\theta_1)^2(2 - \cos\theta_2)^2} \right] \tag{6.2}$$

where r_0 is the radius of the electron, $d\Omega_1$ and $d\Omega_2$ are the elements of the solid angles of γ_1 and γ_2, θ_1 and θ_2 are the corresponding scattering angles, and φ_1 and φ_2 are the azimuths of the counters measured at a plane perpendicular to the propagation axis. Following Hanna (1948), the corresponding probability can be expressed as

$$P(\theta_1, \varphi_1, \theta_2, \varphi_2) = K$$
$$\times \left[\frac{((1 - \cos\theta_1)^3 + 2)((1 - \cos\theta_2)^3 + 2)}{(2 - \cos\theta_1)^3(2 - \cos\theta_2)^3} - \frac{(\sin^2\theta_1 \sin^2\theta_2)\cos 2(\varphi_1 - \varphi_2)}{(2 - \cos\theta_1)^2(2 - \cos\theta_2)^2} \right]. \tag{6.3}$$

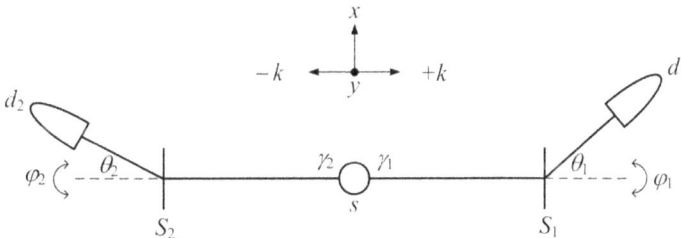

Figure 6.1. Rendition of the original Pryce–Ward experimental diagram. Here, s is the e^+e^- annihilation source of γ_1 and γ_2 emitted in the $+k$ and $-k$ directions (equivalent to the $+z$ and $-z$ directions), respectively. The two linear polarization axes are x and y, which are perpendicular to each other. The angles θ_1 and θ_2 are the scattering angles measured relative to the propagation axis. The azimuths φ_1 and φ_2 are measured at the plane perpendicular to the propagation axis. S_1 and S_2 are the scatterers while d_1 and d_2 are the corresponding detectors.

At $\theta_1 = \theta_2 = \pi/2$, which is convenient for experimental configurations, equation (6.3) reduces to

$$P(\varphi_1, \varphi_2) = \kappa[1 - \cos 2(\varphi_1 - \varphi_2)] \tag{6.4}$$

where $\kappa = (K/16)$. This specific probability expression emphasizes the importance of the $\cos 2(\varphi_1 - \varphi_2)$ term and the Pryce–Ward polarization differential angle

$$\Delta\varphi = 2(\varphi_1 - \varphi_2). \tag{6.5}$$

It is exceedingly interesting to note that the final theoretical result for the scattering quantum entanglement experiment, that is, equation (6.2), is not the focus of this book. Rather, it is an intermediate step, absolutely necessary to arrive at equation (6.2), that gathers the attention of physicists today. In the absence of further comments, that equation is the probability amplitude of two quanta (γ_1 and γ_2) propagating in opposite directions with entangled polarizations in the states $|x\rangle$ and $|y\rangle$ (Ward 1949):

$$|\psi\rangle = (|x_1, y_2\rangle - |y_1, x_1\rangle). \tag{6.6}$$

For some reason, Pryce and Ward chose not to disclose this intermediate step in their 1947 paper, and Ward never bothered to dedicate a paper to this subject, which was *one* of several topics in his doctoral thesis (Ward 1949).

The theoretical result of Pryce and Ward was quickly confirmed by the independent work of American physicists Snyder *et al* (1948) and was also independently reproduced by a young Australian physicist, Richard Henry Dalitz, a recent arrival to England at the time (Ward 2004), who would later become known by his contributions to particle physics including the early research on quarks, the *Dalitz plot*, and *Dalitz pairs*.

6.2.2 Experimental legacy of the Pryce–Ward paper

The three experiments that verified the quantum scattering theory of Pryce and Ward (1947) adopted experimental configurations derived from the schematics disclosed by these authors, which were inspired by Wheeler's written proposal (Wheeler 1946). It is also apparent that since this experimental approach involved scattering prior to detection, it depended on four angular quantities, θ_1, θ_2, φ_1, and φ_2. An experiment in the optical domain, in the absence of scatterers, would only depend on two polarization angles, φ_1 and φ_2, measured on the plane perpendicular to the propagation axis. This modification would require a simplified apparatus as outlined in figure 6.2.

6.3 Ward's doctoral thesis

J C Ward wrote a 47-page doctoral thesis that was divided in two parts and included a total of eight chapters and three appendices. Two of the chapters and two of the appendices in the first part refer to the topic at hand.

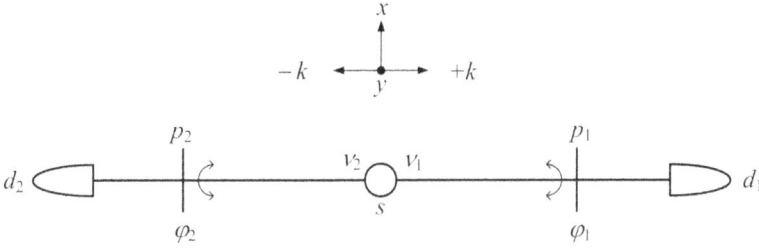

Figure 6.2. Simplified experimental diagram for experiments in the optical domain. Here, s is the photon source of ν_1 and ν_2 emitted in the $+k$ and $-k$ directions (equivalent to the $+z$ and $-z$ directions), respectively. The two linear polarization axes are x and y, which are perpendicular to each other. The angles φ_1 and φ_2 are the polarization angles measured on a plane perpendicular to the propagation axis while d_1 and d_2 are the corresponding detectors.

Ward adopted a heuristic approach, based on conservation of momentum arguments, toward his derivation of the probability amplitude for two counter-propagating quanta, along a common optical axis, with entangled orthogonal polarizations in the states $|x\rangle$ and $|y\rangle$ (Ward 1949). The concept of singlet states, probably adopted from Wheeler's description (Wheeler 1946), was also important. In this regard, Ward initiates his exposition by stating, 'Wheeler proceeded to calculate ... But through the neglect of interference terms he derived an incorrect ... far too small value for the angular correlations ...' (Ward 1949).

In his explanation on the derivation of equation (6.2), Ward wrote, 'it is essential to derive correctly the state vector which properly describes the state of the two quanta' (Ward 1949). Thus, in his own approach Ward began by listing polarization state alternatives for the x and y polarization axes related to two counter-propagating photons:

$$|x, x\rangle, |x, y\rangle, |y, x\rangle, |y, y\rangle. \tag{6.7}$$

After attaching the numeral 1 to the photon associated with the first coordinate x and the numeral 2 to the photon associated with the second coordinate y, these states can be rewritten as

$$|x_1, x_2\rangle, |x_1, y_2\rangle, |y_1, x_2\rangle, |y_1, y_2\rangle. \tag{6.8}$$

The next step for Ward was to consider the momenta options depicted in figure 6.1, and use the same approach to write the states (Ward 1949)

$$|+k, -k\rangle, |-k, +k\rangle. \tag{6.9}$$

Ward then explains that the states for the combined system split into a singlet state and a triplet state. The singlet state is anti-symmetrical both in polarization and momenta (Ward 1949):

$$(|x_1, y_2\rangle - |y_1, x_2\rangle)(|+k, -k\rangle - |-k, +k\rangle). \tag{6.10}$$

Furthermore, the triplet state configuration includes states symmetrical in polarization and momenta. He rejects the option of triplet states since the only admissible option, in the overall configuration, includes quanta with parallel polarizations.

Considering the polarization states exclusively, the probability amplitude including these states is

$$(|x_1, y_2\rangle - |y_1, x_2\rangle) \tag{6.11}$$

and applying the Dirac identities ($|x\rangle|y\rangle = |x, y\rangle$), this can be rewritten as (Duarte 2014)

$$|\psi\rangle = (|x\rangle_1|y\rangle_2 - |y\rangle_1|x\rangle_2)$$

which, following straightforward normalization, becomes

$$|\psi\rangle_- = \frac{1}{\sqrt{2}}(|x\rangle_1|y\rangle_2 - |y\rangle_1|x\rangle_2) \tag{6.12}$$

while its linear combination is

$$|\psi\rangle_+ = \frac{1}{\sqrt{2}}(|x\rangle_1|y\rangle_2 + |y\rangle_1|x\rangle_2). \tag{6.13}$$

These are the iconic equations used in the literature to describe the probability amplitude of entangled polarizations of quanta propagating in opposite directions.

6.4 Summary

Ward's heuristic derivation of $|\psi\rangle = (|x_1, y_2\rangle - |y_1, x_2\rangle)$ relied on the following three physics guidelines.

1. Quanta propagating collinearly, in opposite directions, with equal momenta. This means that $p_1 = +k_1\hbar$ and $p_2 = -k_2\hbar$. Apart from propagation on the same optical axis in opposite directions, $|p_1| = |p_2|$ also means that the wave numbers are equal, $k_1 = k_2$, and subsequently the wavelengths of the quanta are equal, $\lambda_1 = \lambda_2$. This guideline was originally mentioned by Wheeler (1946). This implicitly refers to *indistinguishable quanta*.
2. Only singlet states are allowed.
3. Only combinations with orthogonal polarizations are allowed.

References

Dalitz R H and Duarte F J 2000 John Clive Ward *Phys. Today* **53** 99–100

Duarte F J 2014 *Quantum Optics for Engineers* (New York: CRC)

Hanna R C 1948 Polarization of annihilation radiation *Nature* **162** 332

Pryce M L H and Ward J C 1947 Angular correlation effects with annihilation radiation *Nature* **160** 435

Snyder H S, Pasternack S and Hornbostel J 1948 Angular correlations of scattered annihilated radiation *Phys. Rev.* **73** 440–8

Ward J C 1949 *Some Properties of the Elementary Particles* (Oxford: Oxford University)

Ward J C 2004 *Memoirs of a Theoretical Physicist* (Rochester: Optics Journal)

Wheeler J A 1946 Polyelectrons *Ann. N. Y. Acad. Sci.* **48** 219–38

Chapter 7

The quantum entanglement experiment

Based on the physics elucidated by the Pryce and Ward probability amplitude, $|\psi\rangle = (|x_1, y_2\rangle - |y_1, x_2\rangle)$, the essentials of a quantum entanglement experiment for two quanta propagating in opposite directions with entangled polarization states, $|x\rangle$ and $|y\rangle$, are explained.

7.1 Introduction

Once Pryce and Ward derived

$$|\psi\rangle = (|x\rangle_1|y\rangle_2 - |y\rangle_1|x\rangle_2) \tag{7.1}$$

all the physics had arrived and the bases for the quantum entanglement experiment were set in stone. For completeness, it should be mentioned that, following the usual normalization, equation (7.1) becomes (see, for example, Duarte 2014)

$$|\psi\rangle_- = \frac{1}{\sqrt{2}}(|x\rangle_1|y\rangle_2 - |y\rangle_1|x\rangle_2) \tag{7.2}$$

and its linear combination is

$$|\psi\rangle_+ = \frac{1}{\sqrt{2}}(|x\rangle_1|y\rangle_2 + |y\rangle_1|x\rangle_2). \tag{7.3}$$

7.2 The quantum entanglement experiment

The Pryce–Ward probability amplitude equation describes the following physics.
1. Two quanta are involved: quanta 1 and quanta 2, which for historical reasons can be denoted as γ_1 and γ_2.
2. These two quanta propagate collinearly along the z-axis in opposite directions.

3. Although these quanta are labeled γ_1 and γ_2, they are indistinguishable in the frequency domain.
4. The term $|x\rangle_1|y\rangle_2$ refers to γ_1 in the $|x\rangle$ state and γ_2 in the $|y\rangle$ state.
5. The term $|x\rangle_2|y\rangle_1$ refers to γ_1 in the $|y\rangle$ state and γ_2 in the $|x\rangle$ state.
6. If γ_1 is propagating along the $+z$-axis and is linearly polarized in the $|x\rangle$ state, then γ_2 propagates in the opposite direction $(-z)$ and is polarized in the $|y\rangle$ state. The two polarizations are orthogonal to each other (see figure 7.1).
7. If γ_1 propagates along the $+z$-axis and is linearly polarized in the $|y\rangle$ state, then γ_2 propagates in the opposite direction $(-z)$ and is polarized in the $|x\rangle$ state. The two polarizations are orthogonal to each other (see figure 7.2).

Figures 7.1 and 7.2 are straightforward representations of the probability amplitude given in equation (7.1). Neither polarization analyzers nor detectors are included in the figures even though these elements are obviously a necessary part of the apparatus in order to verify the polarization via measurements.

The following point needs to be reemphasized: all the physics of quantum entanglement is contained in

$$|\psi\rangle_- = \frac{1}{\sqrt{2}}(|x\rangle_1|y\rangle_2 - |y\rangle_1|x\rangle_2).$$

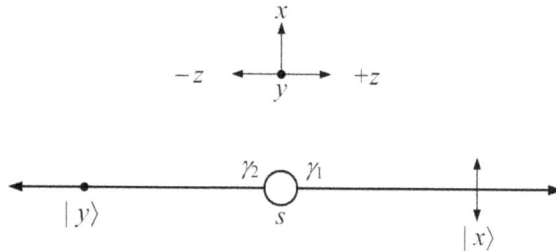

Figure 7.1. Direct quantum entanglement diagram representing the Pryce–Ward probability amplitude $|\psi\rangle = (|x\rangle_1|y\rangle_2 - |y\rangle_1|x\rangle_2)$. Here, γ_1 is depicted as being polarized in the $|x\rangle$ state, which means that γ_2 is polarized in the $|y\rangle$ state. In previous diagrams, the propagation axis z has been depicted with the labels $+k$ and $-k$ to highlight the conservation of momentum.

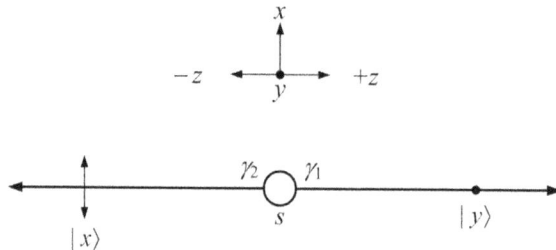

Figure 7.2. Direct quantum entanglement diagram representing the Pryce–Ward probability amplitude $|\psi\rangle = (|x\rangle_1|y\rangle_2 - |y\rangle_1|x\rangle_2)$. Here, γ_1 is depicted as being polarized in the $|y\rangle$ state, which means that γ_2 is polarized in the $|x\rangle$ state. In previous diagrams, the propagation axis z has been depicted with the labels $+k$ and $-k$ to highlight the conservation of momentum.

Polarizers and detectors only serve as necessary, yet secondary, elements to verify the physics of the quantum entanglement probability amplitude. As will be seen in the next chapter, the first experiments and measurements in the field of quantum entanglement were performed using gamma rays. The utilization of gamma rays forced the experimenters to rely on Compton scattering to measure the polarization coincidences.

7.3 Historical notes

The following observations are in order.

1. Although presenting the essence of quantum entanglement, the word entanglement was not used in the Pryce–Ward writings (Pryce and Ward 1947, Ward 1949).
2. The essential features of the schematics of figures 7.1 and 7.2 can be also distilled from the outlines including scattering features disclosed by Pryce and Ward (Pryce and Ward 1947, Ward 1949).
3. As it turns out, historical developments relegated equations (7.1) and (7.2) to the background for a long time, as most of the attention focused on the EPR argument, hidden variable theories, and Bell's theorem.

References

Duarte F J 2014 *Quantum Optics for Engineers* (New York: CRC)

Pryce M L H and Ward J C 1947 Angular correlation effects with annihilation radiation *Nature* **160** 435

Ward J C 1949 *Some Properties of the Elementary Particles* (Oxford: Oxford University Press)

Chapter 8

The annihilation quantum entanglement experiments

The first quantum entanglement experiments (1948–50) utilizing quanta pairs created in the $e^+e^- \to \gamma_1\gamma_2$ annihilation process and Compton scattering detection are described. These experiments yielded reasonable to good agreement with the quantum theory results predicted by Pryce and Ward (1947).

8.1 Introduction

This chapter focuses on the positron–electron annihilation experiments

$$e^+e^- \to \gamma_1\gamma_2 \tag{8.1}$$

performed in the 1948–50 period, in the United States and England, by Bleuler and Bradt (1948), Hanna (1948), and Wu and Shaknov (1950). These were all very brief communications, or letters, contained in one page or less. Two of these papers refer to Wheeler (1946) and all three refer to Pryce and Ward (1947).

Gamma rays demand special propagation and detection optics given their extremely high frequency, $\nu \approx 300 \times 10^{18}$ Hz (or $\lambda \approx 1 \times 10^{-3}$ nm). One additional observation is that the gamma rays emitted in the e^+e^- annihilation, γ_1 and γ_2, are labeled to indicate that they are a pair integrated by two quanta propagating in opposite directions. Given perfect correlation, these quanta have identical, or nearly identical, frequencies so that $\nu_{\gamma_1} = \nu_{\gamma_2}$, or $\nu_{\gamma_1} \approx \nu_{\gamma_2}$. In other words, perfect correlation in the frequency domain would imply that these two quanta are *indistinguishable* in their emission frequency.

8.2 The first three quantum entanglement experiments

Soon after the publication of the Pryce–Ward theory and their experimental schematics in *Nature*, two experimental confirmations followed. The first confirmation was by Bleuler and Bradt (1948), and the second was by Hanna (1948).

doi:10.1088/2053-2563/ab2b33ch8

A third paper on this subject matter, including a more decisive agreement between the polarization–correlation theory and measurements, would be published by Wu and Shaknov (1950).

Common to the three experiments was the use of a Cu^{64} annihilation radiation source, $\lambda \approx 2.42 \times 10^{-3}$ nm (Dumond *et al* 1949), prepared by deuteron radiation. The gamma ray quanta (γ_1 and γ_2) thus produced were incident on aluminum scatterers. In addition to aluminum scatterers, Hanna (1948) also utilized brass scatterers. Bleuler and Bradt (1948) and Hanna (1948) use Geiger counters as detectors. Wu and Shaknov (1950), on the other hand, used scintillation counters known to have an efficiency ten times better than Geiger counters.

All three experiments were slightly different in their measurement configurations. This means that they used the Pryce–Ward equation slightly re-expressed as (Hanna 1948)

$$\sigma(\theta, \varphi) = K \, d\Omega_1 d\Omega_2$$
$$\times \left[\frac{((1 - \cos \theta_1)^3 + 2)((1 - \cos \theta_2)^3 + 2)}{(2 - \cos \theta_1)^3 (2 - \cos \theta_2)^3} - \frac{(\sin^2 \theta_1 \sin^2 \theta_2) \cos 2(\varphi_1 - \varphi_2)}{(2 - \cos \theta_1)^2 (2 - \cos \theta_2)^2} \right] \quad (8.2)$$

to best reproduce their particular angular configurations. As anticipated in chapter 6, at $\theta_1 = \theta_2 = \pi/2$, equation (8.2) reduces to

$$\sigma(\varphi) = \kappa \, d\Omega_1 d\Omega_2 [1 - \cos 2(\varphi_1 - \varphi_2)] \quad (8.3)$$

where $\kappa = (K/16)$. This reduced equation emphasized the importance of the term $\cos 2(\varphi_1 - \varphi_2)$ and the Pryce–Ward polarization differential angle $\Delta \varphi = 2(\varphi_1 - \varphi_2)$.

The aim of these researchers was to measure the coincidence counting rates for orthogonal $C_{|x\rangle}$ and parallel $C_{|y\rangle}$ polarizations. A generic depiction of these experiments is given in figure 8.1.

Table 8.1 tabulates the results reported in these experiments while using aluminum scatterers.

The better agreement between theory and experiment reported by Wu and Shaknov (1950) is most likely due to the significant superiority of scintillation counters over Geiger counters for this class of experiment.

According to Pryce and Ward (1947), the maximum scattering ratio

$$\frac{C_{|x\rangle}}{C_{|y\rangle}} = 2.82 \quad (8.4)$$

occurs at $\theta_1 = \theta_2 = 82°$, which explains why the experimenters positioned their counters at $\theta_1 \approx \theta_2 \approx 90°$.

It is important to point out that following the publication of these results there was a long silence in the experimental community on this subject. Perhaps an implicit agreement that this issue had been settled and that experimental measurements of

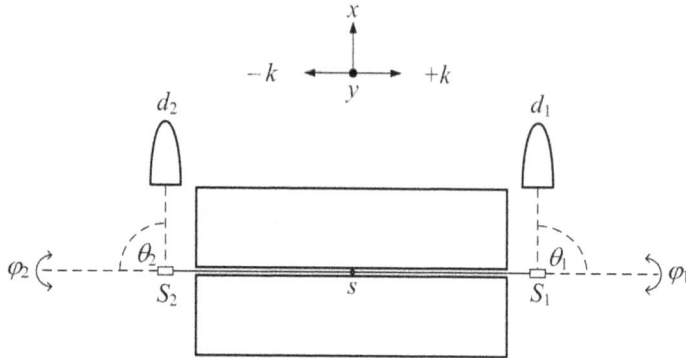

Figure 8.1. Generic schematics for the $e^+e^- \rightarrow \gamma_1\gamma_2$ experiments. Here, s is the Cu^{64} source positioned at the center of the emission channel configured within a Pb block. The two polarization axes are x and y, which are perpendicular to each other. In this diagram, the scattering angles, sustained relative to the propagation axis, are predetermined as $\theta_1 \approx \theta_2 \approx \pi/2$ rads. The azimuths φ_1 and φ_2 are measured at the plane perpendicular to the propagation axis. In this particular figure, $\varphi_1 \approx \varphi_2$, and the measurements are performed as a function of $\Delta\varphi = (\varphi_1 - \varphi_2)$; for instance, $\Delta\varphi = (\varphi_1 - \varphi_2) = \pi/2$ became a preferred angular configuration. S_1 and S_2 are the scatterers while d_1 and d_2 are the corresponding detectors.

Table 8.1. Measured and theoretical coincidence counting rate ratios for orthogonal ($C_{|x\rangle}$) and parallel ($C_{|y\rangle}$) polarizations.

$(C_{|x\rangle}/C_{|y\rangle})_m$	$(C_{|x\rangle}/C_{|y\rangle})_t$	Reference
1.94 ± 0.37	1.70	Bleuler and Bradt (1948)
1.51 ± 0.10	1.86	Hanna (1948)
2.04 ± 0.08	2.00	Wu and Shaknov (1950)

counter-propagating quanta with polarizations orthogonal to each other *did agree* with quantum mechanics. There is more on this in the discussion on interpretations of quantum mechanics (chapters 28 and 29).

One further point is that the three experiments considered here utilized experimental configurations derived from the schematics disclosed by Pryce and Ward (1947), which were inspired by Wheeler's written proposal (Wheeler 1946). It is also obvious that since this experimental approach involved scattering prior to detection, it depended on four angular quantities, θ_1, θ_2, φ_1, and φ_2, as illustrated in figure 8.1. A straightforward experiment in the optical domain and in the absence of scatterers would only depend on two polarization angles, φ_1 and φ_2, and thus would lead to a simplified apparatus as already suggested in chapter 7. The reader should note that in the literature, as time progressed, φ_1 and φ_2 began to be denoted as θ_1 and θ_2 to indicate the polarization angles measured on the plane perpendicular to the propagation axis. A generalized simplified experimental schematic applicable to these observations is outlined in figure 8.2. It should also be mentioned that for maximum correlation, the frequency of the two quanta should be the same, or $\nu_{1_1} = \nu_2$.

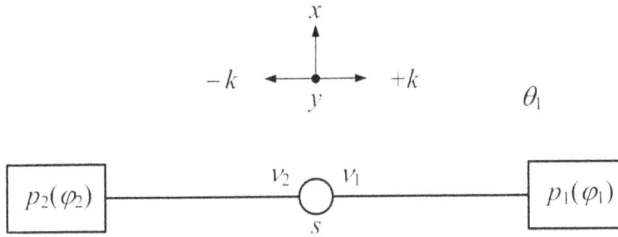

Figure 8.2. Simplified experimental diagram for experiments in the optical domain. Here, s is the photon source of ν_1 and ν_2 ($\nu_1 = \nu_2$) emitted in the $+k$ and $-k$ directions, respectively. The yet undefined polarization-detection portions of the apparatus are free to be positioned as a function of φ_1 and φ_2, which are measured at a plane perpendicular to the propagation axis.

8.3 Further significance of the annihilation experiments

Beyond the experimental significance outlined above, these three homologous experiments had a more profound effect on the physics of quantum entanglement: they pioneered experimental efforts in quantum entanglement measurements and provided an experimental avenue to researchers interested in EPR situations. Indeed, Bohm and Aharonov (1957) include the paper by Wu and Shaknov (1950) in their section entitled 'Experiment verifying the paradox of ERP' (note that due to an apparent typo they wrote ERP rather than EPR). In this paper, Bohm and Aharonov discuss verifying whether the two counter propagating quanta are polarized orthogonal to each other. In this regard, they refer to the Wu and Shaknov (1950) paper as measuring 'essentially for this point, but in a more indirect way' (Bohm and Aharonov 1957). This topic is discussed further in chapter 9.

One interesting caveat relevant to the annihilation experiments is that the first two experiments were largely ignored in the subsequent literature on the subject, and only the Wu and Shaknov experiment continued to be cited. Furthermore, the nature of these citations was unenthusiastic and focused on perceived limitations of the scattering approach. More specifically, it was stated that even though the Wu and Shaknov results were 'compatible with quantum mechanics' the same results had not 'produced evidence against hidden-variable theories' (Clauser *et al* 1969).

References

Bleuler E and Bradt H L 1948 Correlation between the states of polarization of the two quanta of annihilation radiation *Phys. Rev.* **73** 1398

Bohm D and Aharonov Y 1957 Discussion of experimental proof for the paradox of Einstein, Rosen, and Podolsky *Phys. Rev.* **108** 1070–6

Clauser J F, Horner M A, Shimony A and Holt R A 1969 Proposed experiment to test local hidden variable theories *Phys. Rev. Lett.* **23** 880–3

Dumond J W M, Lind D A and Watson B B 1949 Precision measurement of the wavelength and spectral profile of the annihilation radiation from Cu^{64} with the two-meter focusing curved crystal spectrometer *Phys. Rev.* **75** 1226–39

Hanna R C 1948 Polarization of annihilation radiation *Nature* **162** 332

Pryce M L H and Ward J C 1947 Angular correlation effects with annihilation radiation *Nature* **160** 435

Wheeler J A 1946 Polyelectrons *Ann. N. Y. Acad. Sci.* **48** 219–38

Wu C S and Shaknov I 1950 The angular correlation of scattered annihilation radiation *Phys. Rev.* **77** 136

IOP Publishing

Fundamentals of Quantum Entanglement

F J Duarte

Chapter 9

The Bohm and Aharonov paper

The significance of the Bohm and Aharonov (1957) paper in bridging the philosophical path and the physics path of quantum entanglement, by focusing attention on the Wu and Shaknov (1950) paper while also discussing hidden variable theories, is made explicit.

9.1 Introduction

The Bohm and Aharonov (1957) paper played a crucial role in the field of quantum entanglement by presenting together the philosophical arguments developed by Einstein *et al* (1935) and Schrödinger (1935) and the physics developed via the Dirac–Wheeler route (Dirac 1930, Wheeler 1946). However, of the papers cited above Bohm and Aharonov only referred to Einstein *et al* (1935) and did not use the word entanglement.

Bohm and Aharonov's paper was entitled 'Experiment verifying the paradox of ERP' and as its title suggests its main focus was to advance a resolution of what by then had already become known as the *EPR paradox*. Besides a discussion on the philosophy of the situation, Bohm and Aharonov concentrate their attention on the theoretical work of Snyder *et al* (1948) and the experimental results of Wu and Shaknov (1950). The works of Pryce and Ward (1947), Bleuler and Bradt (1948), and Hanna (1948) were not mentioned.

In their paper on spin physics, Bohm and Aharonov (1957) consider a two-atom molecule with a total spin of zero described by a probability amplitude of the form

$$\psi = \frac{1}{\sqrt{2}}\big(\psi_+(1)\psi_-(2) - \psi_-(1)\psi_+(2)\big). \qquad (9.1)$$

Following dissociation of the atoms, a component of the spin of atom 1 is measured: 'because the total spin is still zero, it can immediately be concluded that the same component of the spin of the other particle (2) is opposite to that of 1' (Bohm and Aharonov 1957). These authors go on to indicate that, from their perspective,

doi:10.1088/2053-2563/ab2b33ch9

the experimenter is 'free to choose' even before the measurement has taken place. It should be noted that in the original manuscript, A and B are used to designate the particles rather than 1 and 2. The numbers 1 and 2 are used here since that is the nomenclature used in equation (9.1).

Since the description given above does not fit the classical perspective of measurements on separate systems, these authors advance the notion that 'perhaps … there is some hidden interaction between 2 and 1, or between 2 and the measuring apparatus' (Bohm and Aharonov 1957).

Next they mention that there is an experiment, that of Wu and Shaknov (1950), that shows that their proposed interpretation of the EPR paradox is 'untenable' (Bohm and Aharonov 1957).

As an editing note, it should be mentioned that what Bohm and Aharonov call 'ERP' is in fact EPR, that is, Einstein, Podolsky, and Rosen.

9.2 Significance to the development of quantum entanglement research

The paper of Bohm and Aharonov (1957) was significant in that it, for the first time, brought together the philosophical perspective on quantum entanglement, as articulated by Einstein *et al* (1935) and Schrödinger (1935) and the quantum mechanics of the effect as presented by Snyder *et al* (1948) and Wu and Shaknov (1950). In this regard, they refer to probability amplitudes in the form of equation (9.1) and probability amplitudes of the forms

$$\phi_1 = \frac{1}{\sqrt{2}}(\psi_1 - \psi_2) \qquad (9.2)$$

and

$$\phi_1 = \frac{1}{\sqrt{2}}\left(C_1^x C_2^y - C_1^y C_2^x\right). \qquad (9.3)$$

On the experimental side, Bohm and Aharonov refer to the EPR paradox and the Wu and Shaknov (1950) experiment in their abstract, 'there already is an experiment whose significance in regard to this problem has not yet been explicitly brought out', only to moderate their stance later by expressing that Wu and Shaknov provided relevant measurements 'essentially for this point, but in a more indirect way' (Bohm and Aharonov 1957). Their discussion of the Wu and Shaknov experiment went further, and in the only experimental diagram of their paper a reduced and stylized version of the Compton scattering experimental configuration is provided.

What Bohm and Aharonov did not mention was the fact that the Ward probability amplitude (Ward 1949) is an essential step toward the derivation of the cross-section equation for scattering published by Pryce and Ward (1947) and utilized by Wu and Shaknov in their comparison of their polarization coincidence measurements with theory. If the measurements are done properly, then they are a correct, albeit 'indirect,' validation of

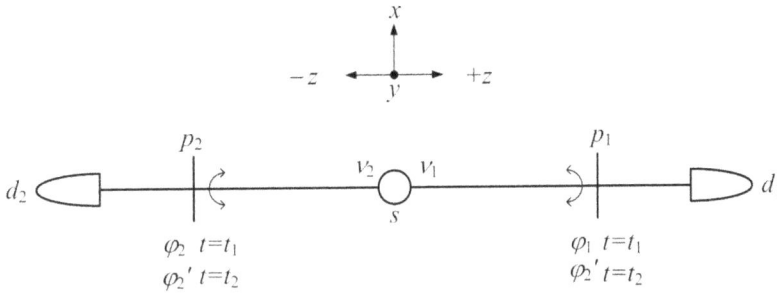

Figure 9.1. Rendition of the thought experiment suggested by Bohm and Aharonov (1957). At $t = t_1$, φ_1 and φ_2 are at their initial settings. At $t = t_2$, where $t > t_2$, $\varphi_1 = \varphi_1'$ and $\varphi_2 = \varphi_2'$.

$$|\psi\rangle = \big(|x_1, y_2\rangle - |y_1, x_2\rangle\big). \tag{9.4}$$

In his explanation on the derivation of

$$\frac{r_0^4 d\Omega_1 d\Omega_2}{16}$$
$$\times \left[\frac{((1 - \cos\theta_1)^3 + 2)((1 - \cos\theta_2)^3 + 2)}{(2 - \cos\theta_1)^3(2 - \cos\theta_2)^3} - \frac{(\sin^2\theta_1 \sin^2\theta_2)\cos 2(\varphi_1 - \varphi_2)}{(2 - \cos\theta_1)^2(2 - \cos\theta_2)^2} \right] \tag{9.5}$$

Ward wrote: 'it is essential to derive correctly the state vector which properly describes the state of the two quanta' (Ward 1949). That state vector is given in equation (9.4).

One further contribution of the Bohm and Aharonov paper is the suggestion to alter the settings of the measuring apparatus 'even while the atoms are still in flight' (see figure 9.1).

9.3 Philosophy and physics

Bohm and Aharonov (1957) succeeded in shinning a light on the fact that proposed resolutions of the EPR paradox invoking 'hidden interactions' (see, for example, Bohm 1952) were at odds with 'current quantum theory'. Indeed, John S Bell (1964) would focus his attention on this very Bohm and Aharonov paper while deriving his celebrated Bell inequalities.

For the sake of completeness, it should also be mentioned that Bohm in his book *Quantum Theory* (Bohm 1951) did advance a direct one-page argument entitled 'Proof that quantum theory is incompatible with hidden variables'. In it, Bohm suggests that hidden variables are incompatible with Heisenberg's uncertainty principle.

References

Bell J S 1964 On the Einstein–Podolsky–Rosen paradox *Physics* **1** 195–200

Bleuler E and Bradt H L 1948 Correlation between the states of polarization of the two quanta of annihilation radiation *Phys. Rev.* **73** 1398

Bohm D 1951 *Quantum Theory* (Englewood Cliffs, NJ: Prentice-Hall)

Bohm D 1952 A suggested interpretation of quantum theory using 'hidden' variables. I *Phys. Rev.* **85** 166–79

Bohm D and Aharonov Y 1957 Discussion of experimental proof for the paradox of Einstein, Rosen, and Podolsky *Phys. Rev.* **108** 1070–6

Dirac P A M 1930 On the annihilation of electrons and protons *Math. Proc. Camb. Philos. Soc.* **2** 361–75

Einstein A, Podolsky B and Rosen N 1935 Can quantum mechanical description of physical reality be considered complete? *Phys. Rev.* **47** 777–80

Hanna R C 1948 Polarization of annihilation radiation *Nature* **162** 332

Pryce M L H and Ward J C 1947 Angular correlation effects with annihilation radiation *Nature* **160** 435

Schrödinger E 1935 Discussion of probability relations between separated systems *Math. Proc. Camb. Philos. Soc.* **31** 555–63

Snyder H S, Pasternack S and Hornbostel J 1948 Angular correlations of scattered annihilated radiation *Phys. Rev.* **73** 440–8

Ward J C 1949 *Some Properties of the Elementary Particles* (Oxford: Oxford University Press)

Wheeler J A 1946 Polyelectrons *Ann. N. Y. Acad. Sci.* **48** 219–38

Wu C S and Shaknov I 1950 The angular correlation of scattered annihilation radiation *Phys. Rev.* **77** 136

Chapter 10

Bell's theorem

Bell's famous theorem is reviewed and presented in a straightforward style using Bell's notation. The significance of Bell's theorem reaffirming the incompatibility of local hidden variable theories with quantum mechanics is highlighted.

10.1 Introduction

John S Bell's paper, entitled 'On the Einstein–Podolsky–Rosen paradox', was a breakthrough in the interpretational arena of quantum mechanics. Bell's paper provided a mathematical mechanism that would show that *hidden variable theories* were incompatible with the predictions of quantum mechanics. That this is the case can be deduced directly from the introduction written by Bell to his paper. In his introduction, Bell refers immediately to Einstein *et al* (1935) and their claim that quantum mechanics was incomplete and should be 'supplemented by additional variables'. He then continues with, 'that idea will be formulated mathematically and shown to be incompatible with the statistical predictions of quantum mechanics' (Bell 1964). This mathematical statement became known as either *Bell's theorem* or *Bell's inequality*.

He also mentions in his introduction that a hidden variable interpretation of quantum mechanics had already been formulated by Bohm (1952). In the next section of his paper he refers to the Bohm and Aharonov (1957) paper as providing an outline to test the EPR argument in a quantum mechanical framework.

Two observations are in order here: as already exposed in the previous chapters, the outline considered by Bohm and Aharonov (1957) did include knowledge from the 'physics path' history; and Bell did not yet mention the word *entanglement* introduced by Schrödinger (1935).

10.2 von Neumann's work

As a preamble to a discussion on Bell's work it should be mentioned that as early as 1932 John von Neumann wrote about quantum mechanics and hidden variables.

10-1

In his book entitled *Mathematical Foundations of Quantum Mechanics*, in its preface von Neumann alerts the reader that 'hidden parameters' are *incompatible* with the fundamental postulates of quantum mechanics. About three-quarters into the text, in a section entitled 'Proof of statistical formulas', he tackles the issue of hidden variables head-on and writes, 'It is therefore not, as is often assumed, a question of re-interpretation of quantum mechanics, the present system of quantum mechanics would have to be objectively false, in order that another description of the elementary processes than the statistical one be possible' (von Neumann 1955). However, Bell considered von Neumann's proof 'wanting' (Bell 1964), and proceeded to offer his own.

It should be added that von Neumann indicates in his preface that his book was, in part, motivated to provide a description of quantum mechanics different from that of Dirac and based on mathematical rigor. In this regard, he was referring to Dirac's papers and to the first edition of Dirac's book, which was published in 1930 (Dirac 1978). Although von Neumann criticized some aspects of Dirac's mathematical approach, he used two words to sum up Dirac's book: *brevity* and *elegance*.

10.3 Bell's theorem or Bell's inequalities

Bell begins his exposition by recasting in his own words the quantum mechanical ideas first pioneered by Wheeler (1946), although he does not reference Wheeler. For consistency and clarity, Wheeler's statement is reproduced here: 'According to the pair theory, if one of these photons is linearly polarized in one plane, then the photon that goes off in the opposite direction with equal momentum is linearly polarized in the perpendicular plane'. Bell begins to add his own originality to the argument by adding, 'it follows that the result of any such measurement must actually be predetermined ... this predetermination implies the possibility of a more complete specification of the state' (Bell 1964).

Here, a succinct exposition of his theorem is given using a slightly different notation to the original work.

Bell defines a hidden variable λ, or hidden variables, as having a probability density $\rho(\lambda)$ such that its normalized probability is

$$\int \rho(\lambda)d\lambda = 1. \tag{10.1}$$

Then, the correlation between observables, which are functions of this hidden variable, $A(x, \lambda)$ and $B(y, \lambda)$, is stated by Bell as

$$P(x, y) = \int A(x, \lambda)B(y, \lambda)\rho(\lambda)d\lambda. \tag{10.2}$$

x and y are directions, and the condition of locality requires that A depends only on x and B depends only on y. The variables x' and y' are assumed to be alternative settings in the two measuring instruments. Following the format established in equation (10.2),

$$|P(x, y) - P(x, y')| \leqslant \int |[A(x, \lambda)B(y, \lambda) - A(x, \lambda)B(y', \lambda)]| \rho(\lambda)d\lambda \qquad (10.3)$$

and

$$|P(x', y') + P(x', y)| \leqslant \int |[A(x', \lambda)B(y', \lambda) - A(x', \lambda)B(y, \lambda)]| \rho(\lambda)d\lambda. \qquad (10.4)$$

Since $|A(x, \lambda)| \leqslant 1$, from equation (10.3),

$$|P(x, y) - P(x, y')| \leqslant \int |[B(y, \lambda) - B(y', \lambda)]| \rho(\lambda)d\lambda \qquad (10.5)$$

and similarly,

$$|P(x', y') + P(x', y)| \leqslant \int |[B(y', \lambda) + B(y, \lambda)]| \rho(\lambda)d\lambda. \qquad (10.6)$$

Adding equations (10.5) and (10.6),

$$\begin{aligned}
|P(x, y) &- P(x, y')| + |P(x', y') + P(x', y)| \\
&\leqslant \int [|B(y, \lambda) - B(y', \lambda)| + |B(y', \lambda) + B(y, \lambda)|] \rho(\lambda)d\lambda
\end{aligned} \qquad (10.7)$$

and given the definitions $|B(y, \lambda)| = \pm 1$, $|B(y', \lambda)| = \pm 1$,

$$\int [|B(y, \lambda) - B(y', \lambda)| + |B(y', \lambda) + B(y, \lambda)|] \rho(\lambda)d\lambda = 2 \qquad (10.8)$$

leads directly to Bell's inequality, which is also known as Bell's theorem

$$|P(x, y) - P(x, y')| + |P(x', y') + P(x', y)| \leqslant 2. \qquad (10.9)$$

Using Pryce–Ward probability amplitudes, of the form $|\psi\rangle = (|x\rangle_1 |y\rangle_2 - |y\rangle_1 |x\rangle_2)$ (Pryce and Ward 1947, Ward 1949), introduced in chapter 6, to calculate the corresponding quantum entanglement probabilities for the alternatives $P(x, y)$, $P(x', y')$, $P(x', y)$, and $P(x, y')$, leads to numerical values for the sum

$$\Sigma_P = |P(x, y) - P(x, y')| + |P(x', y') + P(x, y')| \qquad (10.10)$$

greater than 2. Hence, for $\Sigma_P \geqslant 2$, Bell's inequality given in equation (10.9) is violated. Bell's theorem is outlined pictorially in figure 10.1.

10.4 An additional perspective on Bell's theorem

A slightly different perspective might help to further illustrate Bell's theorem. A main assumption in this theorem is the locality postulate. This assumption means that for two separated apparatus the experimental settings in one apparatus (A) have no effect on the settings on the other apparatus (B). If these settings are x, y, x', and y', then, *if* the postulate of locality holds, the inequality

$$|P(x, y) - P(x, y')| + |P(x', y') + P(x', y)| \leqslant 2$$

is true. On the other hand, if locality does not hold, then Bell's inequality is violated.

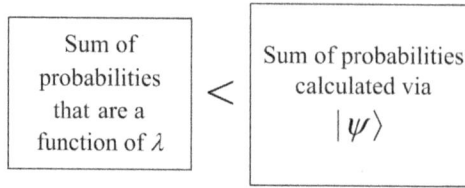

Figure 10.1. The sum of ordinary probabilities that are a function of a hidden variable λ have a maximum value of 2 (Bell 1964). The corresponding probabilistic sum where the probabilities are quantum entanglement probabilities, calculated from probability amplitudes of the form $|\psi\rangle = (|x\rangle_1|y\rangle_2 - |y\rangle_1|x\rangle_2)$, have a numerical value greater than 2.

10.5 Example

As hinted in chapter 6 and as will be shown in chapter 21, the quantum entanglement probability applicable to two quanta propagating in opposite directions while exhibiting orthogonal polarizations is given by

$$P(\varphi_1, \varphi_2) = -\cos 2(\varphi_1 - \varphi_2) \qquad (10.11)$$

for $\varphi_1 = 0$, $\varphi_2 = \pi/3$, $\varphi_1' = \pi/6$, and $\varphi_2' = 0$ (Duarte, 2014), and

$$\Sigma_P = |P(\varphi_1, \varphi_2) - P(\varphi_1, \varphi_2')| + |P(\varphi_1', \varphi_2') + P(\varphi_1', \varphi_2)|$$

leads to the numerical expression

$$\Sigma_P = |{+}0.5 + 1.0| + |{-}0.5 - 0.5| \qquad (10.12)$$

$$\Sigma_P = 2.5 \qquad (10.13)$$

so that

$$\Sigma_P \geqslant 2. \qquad (10.14)$$

This shows that Bell's inequality is violated when using probabilities derived from the Pryce–Ward probability amplitude for quantum entanglement.

10.6 More philosophy and physics

The impact of Bell's theorem on the physics community interested in a deterministic interpretation of quantum mechanics was enormous. Bell's work neatly showed that the notion of hidden variables and quantum mechanics were incompatible. This topic will be revisited in the section on the interpretation of quantum mechanics.

A utilitarian contribution of Bell's theorem has been its use in the field of quantum cryptography, where Bell's inequality is applied to determine if the transmission and reception of quanta pairs is free from third-party intrusion (see chapters 21 and 22).

References

Bell J S 1964 On the Einstein–Podolsky–Rosen paradox *Physics* **1** 195–200

Bohm D 1952 A suggested interpretation of quantum theory using 'hidden' variables. I *Phys. Rev.* **85** 166–79

Bohm D and Aharonov Y 1957 Discussion of experimental proof for the paradox of Einstein, Rosen, and Podolsky *Phys. Rev.* **108** 1070–6

Dirac P A M 1978 *The Principles of Quantum Mechanics* 4th edn (Oxford: Oxford University Press)

Duarte F J 2014 *Quantum Optics for Engineering* (New York: CRC)

Einstein A, Podolsky B and Rosen N 1935 Can quantum mechanical description of physical reality be considered complete? *Phys. Rev.* **47** 777–80

Pryce M L H and Ward J C 1947 Angular correlation effects with annihilation radiation *Nature* **160** 435

Schrödinger E 1935 Discussion of probability relations between separated systems *Math. Proc. Camb. Philos. Soc.* **31** 555–63

von Neumann J 1955 *Mathematical Foundations of Quantum Mechanics* (Princeton, NJ: Princeton University Press) Note: this is a translation of the original title published in German by Springer in 1932

Ward J C 1949 *Some Properties of the Elementary Particles* (Oxford: Oxford University Press)

Wheeler J A 1946 Polyelectrons *Ann. N. Y. Acad. Sci.* **48** 219–38

Chapter 11

Feynman's Hamiltonians

Feynman's derivation, in 1965, of $|s\rangle = (|B\rangle - |A\rangle)$ just a few steps from $|\psi\rangle = (|x_1, y_2\rangle - |y_1, x_2\rangle)$ is reviewed.

11.1 Introduction

The heuristic route to the probability amplitude for quantum entanglement, that is, the Ward approach (Ward 1949), has already been described in chapter 6. There are two other approaches. One of these approaches is examined here and follows Feynman's discussion of differential equations including the Hamiltonian matrix H_{ij} (Feynman *et al* 1965). Although Feynman did not specifically derive $|\psi\rangle = (|x_1, y_2\rangle - |y_1, x_2\rangle)$, he left things advanced enough in his discussion of two-state systems so that the desired probability amplitude could be derived with further steps utilizing some of Dirac's identities (Duarte 2014).

11.2 Probability amplitudes via Hamiltonians *à la Feynman*

The following description on two-state systems follows the style adopted by Duarte (2014) and is based on the physics presented by Feynman in the now famous *Feynman's Lectures on Physics* (Feynman *et al* 1965). In his description, Feynman begins with the differential equation describing the dynamics for a two-state system (see figure 11.1)

$$i\hbar \frac{dC_i}{dt} = \sum H_{ij} C_j \tag{11.1}$$

where H_{ij} is the Hamiltonian. Using the Dirac principle

$$\langle \psi | \phi \rangle = \sum_i \langle \psi | i \rangle \langle i | \phi \rangle \tag{11.2}$$

doi:10.1088/2053-2563/ab2b33ch11

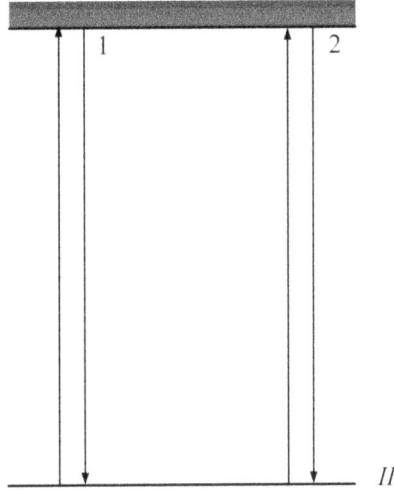

Figure 11.1. Generic two-state system where transitions from a given state $|II\rangle$ to two other closely lying, and overlapping, states $|1\rangle$ and $|2\rangle$ form part of an interferometric system.

and setting $\phi = \psi = II$,

$$\langle II|II\rangle = \langle II|1\rangle\langle 1|II\rangle + \langle II|2\rangle\langle 2|II\rangle \tag{11.3}$$

$$\langle II|II\rangle = \langle II|1\rangle\langle II|1\rangle^* + \langle II|2\rangle\langle II|2\rangle^* \tag{11.4}$$

$$\langle II|II\rangle = C_I C_I{}^* + C_{II} C_{II}{}^* \tag{11.5}$$

$$\langle II|II\rangle = |C_I|^2 + |C_{II}|^2 \tag{11.6}$$

and since

$$\langle II|II\rangle = |C_I|^2 + |C_{II}|^2 = 1 \tag{11.7}$$

we have

$$C_{II} = \frac{1}{\sqrt{2}}(C_1 + C_2) \tag{11.8}$$

$$C_I = \frac{1}{\sqrt{2}}(C_1 - C_2). \tag{11.9}$$

Given that $C_1 = \langle 1|\phi\rangle$ and $C_2 = \langle 2|\phi\rangle$, for a two-state system

$$\langle II|\phi\rangle = \frac{1}{\sqrt{2}}(\langle 1|\phi\rangle + \langle 2|\phi\rangle) \tag{11.10}$$

$$\langle I|\phi\rangle = \frac{1}{\sqrt{2}}(\langle 1|\phi\rangle - \langle 2|\phi\rangle). \tag{11.11}$$

Following the abstraction of $|\phi\rangle$, equation (11.11) reduces to

$$\langle I| = \frac{1}{\sqrt{2}}(\langle 1| - \langle 2|) \tag{11.12}$$

which is equivalent to

$$|I\rangle = \frac{1}{\sqrt{2}}(|1\rangle - |2\rangle). \tag{11.13}$$

It also follows that

$$|II\rangle = \frac{1}{\sqrt{2}}(|1\rangle + |2\rangle). \tag{11.14}$$

This is where Feynman left it in 1965 in his discussion of the dynamics of a two-state system (Feynman *et al* 1965).

11.3 Arrival to quantum entanglement probability amplitudes

The probability amplitude *à la Feynman* expressed via equation (11.13) can be rewritten as

$$|s\rangle = \frac{1}{\sqrt{2}}(|B\rangle - |A\rangle). \tag{11.15}$$

For an ensemble of particles, Dirac (1978) introduces identities that lead to (Duarte 2014)

$$|B\rangle = |x\rangle_1 |y\rangle_2 \tag{11.16}$$

$$|A\rangle = |y\rangle_1 |x\rangle_2 \tag{11.17}$$

so that equation (11.15) becomes (Duarte 2014)

$$|s\rangle_- = \frac{1}{\sqrt{2}}(|x\rangle_1 |y\rangle_2 - |y\rangle_1 |x\rangle_2) \tag{11.18}$$

where the negative sign in $|s\rangle_-$ refers to the sign on the right-hand side. This is the probability amplitude.

11.4 Discussion

The arrival to the equation for the probability amplitude of two counter propagating quanta, with entangled polarization states $|x\rangle$ and $|y\rangle$, equation (11.18), via the Feynman route is rather abstract and does not illustrate the underlying physics. This topic will be revisited from an interferometric perspective in chapter 17.

References

Dirac P A M 1978 *The Principles of Quantum Mechanics* 4th edn (Oxford: Oxford University Press)

Duarte F J 2014 *Quantum Optics for Engineers* (New York: CRC)

Feynman R P, Leighton R B and Sands M 1965 *The Feynman Lectures on Physics* vol III (Reading, MA: Addison-Wesley)

Ward J C 1949 *Some Properties of the Elementary Particles* (Oxford: Oxford University Press)

Chapter 12

The second Wu quantum entanglement experiment

Wu's second experiment on quantum entanglement using $e^+e^- \rightarrow \gamma_1\gamma_2$ annihilation, once again reaffirming the Pryce and Ward probability amplitude, and in light of Bell's theorem, is described and discussed.

12.1 Introduction

In chapter 8 the annihilation $e^+e^- \rightarrow \gamma_1\gamma_2$ quantum entanglement experiments of Bleuler and Bradt (1948), Hanna (1948), and Wu and Shaknov (1950) were considered and described. In this chapter, attention is focused on the second annihilation experiment conducted by Wu and colleagues (Kasday *et al* 1975).

12.2 Salient features

Following the publication of Bell's theorem (Bell 1964), Wu and colleagues revisited the positron–electron annihilation experiments (Kasday *et al* 1975). In these writings Wu and co-authors address directly the relevance of their measurements to the EPR argument (Einstein *et al* 1935), an exercise that was not done while discussing the original experiment (Wu and Shaknov 1950). In this regard they refer to 'the belief of Einstein and others that it is possible to find a theory which provides more than the statistical predictions of quantum mechanics' (Kasday *et al* 1975).

These authors continue to explain that Compton scattering, for gamma rays, is analogous to a linear-polarization analyzer (Kasday *et al* 1975), as already hinted in chapter 8.

Wu and coworkers then refer to the pre-scattering probability amplitude

$$|\psi\rangle = \frac{1}{\sqrt{2}}(|x\rangle_1|y\rangle_2 - |y\rangle_1|x\rangle_2) \qquad (12.1)$$

doi:10.1088/2053-2563/ab2b33ch12

'first worked out by Pryce and Ward (1947)' (Kasday *et al* 1975) as being essential to analyze their experimental measurements.

Next, reference is made of Bohm and Aharonov (1957) expressing that the previous scattering measurements in the annihilation experiment (Wu and Shaknov 1950) 'were sufficient to rule out certain hypothetical modifications of quantum mechanics' derived from the EPR argument (Kasday *et al* 1975).

These authors devote several pages to thoroughly describing their new and improved experimental apparatus (see figure 12.1) and scattering measurements. The agreement between their scattering measurements and quantum theory, via equation (12.1), is within a margin of error of ~2.1%.

12.3 Bell's theorem and hidden variables

Kasday *et al* (1975) dedicate their discussion to the relevance of Bell's theorem to the Compton scattering configuration. In this regard, relying partly on third-party references, they conclude that their scattering configuration is not suitable for a rigorous test of Bell's theorem. However, they also specifically articulate that their $e^+e^- \rightarrow \gamma_1\gamma_2$ quantum entanglement experiment 'are thus evidence against local hidden-variable theories' (Kasday *et al* 1975).

To question the claim of the scattering experiments as *the first observations of quantum entanglement* is to question the impact of all other physics experiments that rely on more than one direct and immediate observation to deduce a valid result.

Here is the essence of what is known about the scattering experiments:

1. The scattering angle was set at $\theta \approx \pi/2$, near the optimum angle $\theta = 82°$ suggested by theory (Pryce and Ward 1947).
2. The electron in the scatterer vibrates along the direction of the polarization axis of the incident quanta. The direction of scattering is perpendicular to the polarization axis of the incident quanta (Kasday *et al* 1975).
3. After the scattering angle was set at $\theta \approx \pi/2$, the detectors were positioned at the azimuths φ_1 and φ_2 as illustrated in figure 12.1.

This scenario is not an overly complicated multiple-step process. It is a two-stage measurement process rather than a single-stage measurement process. According to van Kampen's theorems on quantum measurements, 'quantum mechanics is concerned with macroscopic phenomena, which are not perturbed by observation' (van Kampen 1988). As long as no observational attempts are made to intrude in the measurement process consisting of the incidence of the entangled quanta and the scattering of the radiation toward the detector, there is no pragmatic motive to doubt the validity of the measurements from the scattering experiments.

Furthermore, all the polarization coincidence results under consideration, i.e. those of Hanna (1948), Bleuler and Bradt (1948), Wu and Shaknov (1950), and Kasday *et al* (1975), agree fairly well with the quantum mechanical predictions, via $|\psi\rangle = (|x\rangle_1|y\rangle_2 - |y\rangle_1|x\rangle_2)$, made by Pryce and Ward (1947).

The second Wu experiment was not the only experiment to revisit the annihilation measurements. An additional experiment was also performed by Wilson *et al* (1976).

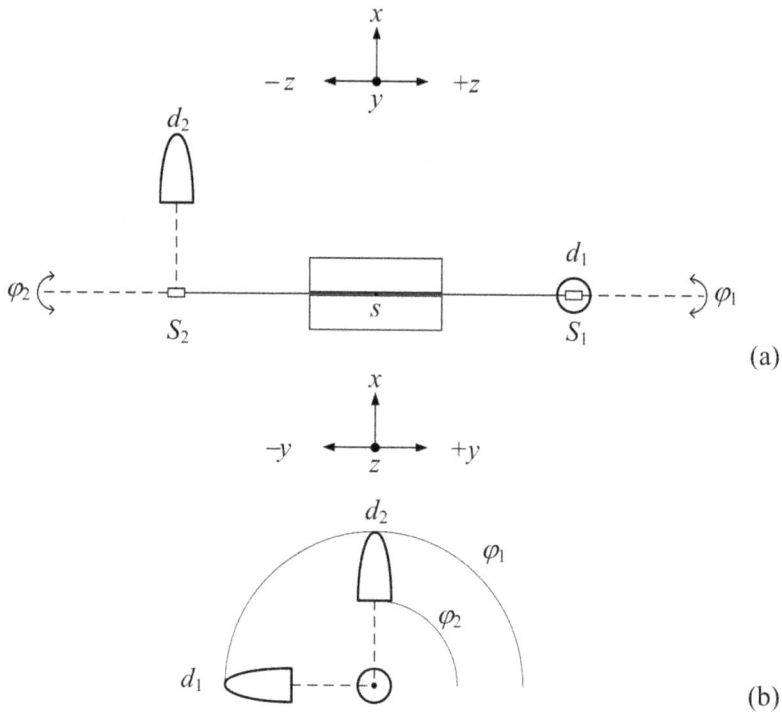

Figure 12.1. A simplified rendition of one of the experimental configurations considered in the second Wu experiment. Here the Pryce–Ward polarization differential angle is $(\varphi_2 - \varphi_1) = \pi/2$. The circular surface area of the detector d_1, deployed in the $+z$ direction, faces directly toward the scatterer.

These researchers used the optimum angle of scattering $\theta = 82°$ (Pryce and Ward 1947) and their conclusion was as follows: 'there appears to be no evidence for spontaneous localization of the quanta and, irrespective of their separation, their properties are accurately described by a pure state' (Wilson *et al* 1976). By a 'pure state' they mean equation (12.1).

References

Bell J S 1964 On the Einstein–Podolsky–Rosen paradox *Physics* **1** 195–200

Bleuler E and Bradt H L 1948 Correlation between the states of polarization of the two quanta of annihilation radiation *Phys. Rev.* **73** 1398

Bohm D and Aharonov Y 1957 Discussion of experimental proof for the paradox of Einstein, Rosen, and Podolsky *Phys. Rev.* **108** 1070–6

Einstein A, Podolsky B and Rosen N 1935 Can quantum mechanical description of physical reality be considered complete? *Phys. Rev.* **47** 777–80

Hanna R C 1948 Polarization of annihilation radiation *Nature* **162** 332

Kasday L R, Ullman J D and Wu C S 1975 Angular correlation of Compton-scattered annihilation photons and hidden variables *Il Nuovo Cimento* **25** 633–61

Pryce M L H and Ward J C 1947 Angular correlation effects with annihilation radiation *Nature* **160** 435

van Kampen N G 1988 Ten theorems about quantum mechanical measurements *Phys.* A **153** 97–113

Wilson A R, Lowe J and Butt D K 1976 Measurement of the relative planes of polarization of annihilation quanta as a function of separation distance *J. Phys. G: Nucl. Phys.* **2** 613–24

Wu C S and Shaknov I 1950 The angular correlation of scattered annihilation radiation *Phys. Rev.* **77** 136

Chapter 13

The hidden variable theory experiments

Experimental attempts to test for local hidden variables in light of Bell's theorem are described. These experimental efforts introduced the optically pumped ^{40}Ca source that emitted quanta pairs, with correlated polarizations, in the visible spectrum. Neither the word *entanglement*, nor $|\psi\rangle = (|x\rangle_1 |y\rangle_2 - |y\rangle_1 |x\rangle_2)$, were mentioned in the resulting publications.

13.1 Introduction

The annihilation $e^+e^- \rightarrow \gamma_1\gamma_2$ quantum entanglement experiments of Bleuler and Bradt (1948), Hanna (1948), and Wu and Shaknov (1950) are described in chapter 8. In this chapter, and in chapter 14, attention is focused on a series of second-generation quantum entanglement experiments: the optical experiments.

The experiments considered here were motivated by the notion that the annihilation experiments with gamma rays were not sufficient to rule out hidden variable theories. Thus began a series of experiments, in the optical domain, based on the writings of Bohm and Aharonov (1957). These experiments are mentioned, and some are further described, here.

13.2 Testing for local hidden variable theories

In a paper entitled 'Proposed experiment to test local hidden variable theories' Clauser *et al* (1969) wrote in reference to the work of Wu and Shaknov (1950): 'Although the polarization state of the pair is suitable ... their high energy requires the use of Compton polarimeters. Thus, instead of directly examining the polarization correlations, Wu and Shaknov examined the polarization-dependent joint distribution for Compton scattering of the pair'. This paper was followed by a paper entitled 'Experimental consequences of objective local theories' (Clauser and Horne 1974) in which Bell inequalities in a generalized form, applicable to optical experiments, were produced:

$$-1 \leqslant R(\varphi_1, \varphi_2) - R(\varphi_1, \varphi_2')$$
$$+ R(\varphi_1', \varphi_2') + R(\varphi_1', \varphi_2) - R_1(\varphi_1') - R_2(\varphi_2) \leqslant 0. \tag{13.1}$$

In this inequality $R(\varphi_1, \varphi_2)$, and so on, are coincidence rates with the polarizers oriented at certain angles such as φ_1, φ_2, φ_1', and φ_2'. For instance, the $R(\varphi_1, \varphi_2)$ coincidence rate is measured with the first polarizer at an angular position φ_1 and the second polarized at an angular position φ_2, and so on. $R_1(\varphi_1')$ is the measured rate with the second polarizer absent and the first polarized set at φ_1'. $R_2(\varphi_2)$ is the measured rate with the first polarizer absent and the second polarized set at φ_2. The φ_1 and φ_2 angular orientations are illustrated in figure 13.1. The reader will recall that the sum

$$R(\varphi_1, \varphi_2) - R(\varphi_1, \varphi_2') + R(\varphi_1', \varphi_2') + R(\varphi_1', \varphi_2) \tag{13.2}$$

is a slightly modified expression for the left-hand side of Bell's inequality, or

$$|P(x, y) - P(x, y')| + |P(x', y') + P(x', y)| \leqslant 2. \tag{13.3}$$

The quantum mechanical result predicted by Clauser *et al* (1969) is

$$\frac{R(\varphi)}{R_0} = \varepsilon\, F(\theta)\cos 2\varphi \tag{13.4}$$

where R_0 is the coincidence rate measured in the absence of polarizers, ε is a factor depending on the efficiency of the polarizers, $F(\theta)$ is a geometrical function, and φ is the angle 'between the polarizer axes' (Clauser *et al* 1969). Here, the reader should remember that in chapter 6 the Pryce–Ward polarization differential angle was defined as $\Delta\varphi = 2(\varphi_1 - \varphi_2)$.

13.3 Early optical experiment

Clauser *et al* (1969) also proposed to modify an experiment reported by Kocher and Commins (1967), in which a H_2 arc lamp was used to produce UV excitation leading to the population of an upper-lying level in ^{40}Ca that results in the emission of two-step sequential transitions $4p^{2\,1}S_0 - 4p4s^1P_1$ ($\lambda_1 = 551.3$ nm) and $4p4s^1P_1 - 4s^{2\,1}S_0$ ($\lambda_2 = 422.7$ nm) with correlated polarizations (see figure 13.2). The experimental configuration of this optical experiment consists simply of an optical axis with the source at the center and a detector at each end of the optical axis and preceded by a polarization analyzer (see figure 13.3). This appears to be the initial stage for optical versions of the Wheeler–Pryce–Ward quantum entanglement configuration. In this regard, Clauser *et al* (1969) include an interesting line in their paper: 'neither of the experimental realizations (Wu and Shaknov 1950, Kocher and Commins 1967) of Bohm's gedankenexperiment has produced evidence against local hidden-variable theories, even though the results of both are compatible with quantum mechanical predictions'. The reference to Bohm is to his book *Quantum Theory* (Bohm 1951). This statement is not corroborated by the literature since the Wu and Shaknov experiment dates back to 1950 and follows the Wheeler–Pryce–Ward route (Wheeler 1946, Pryce and Ward 1947).

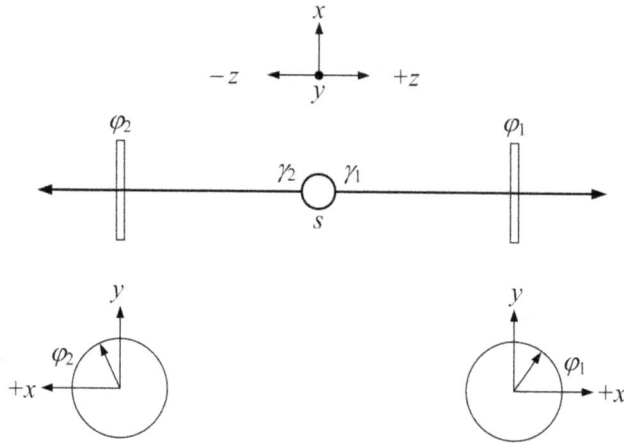

Figure 13.1. Schematics illustrating that the φ_1 and φ_2 are the transmission angles of the polarization analyzers applicable to the Bell-based Clauser and Horne (1974) hidden variable analysis. The circular diagrams shown below depict the polarization analyzers as viewed through their back surface plane that is perpendicular to the propagation axis.

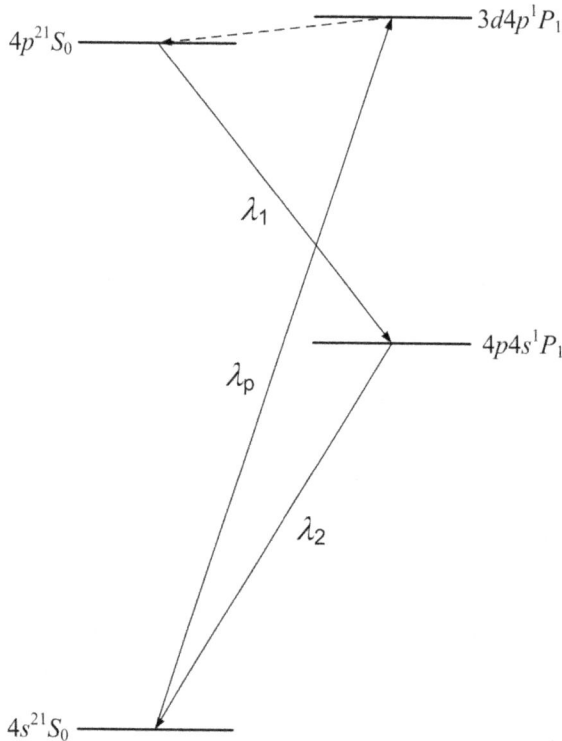

Figure 13.2. Approximate energy level diagram for ^{40}Ca relevant to quantum entanglement experiments. In the original experimental configuration of Kocher and Commins (1967) the excitation wavelength was $\lambda_{UV} = 227.5$ nm. The green transition $4p^{2\,1}S_0 - 4p4s^1P_1$ occurs at $\lambda_1 = 551.3$ nm and the blue transition $4p4s^1P_1 - 4s^{2\,1}S_0$ occurs at $\lambda_2 = 422.7$ nm.

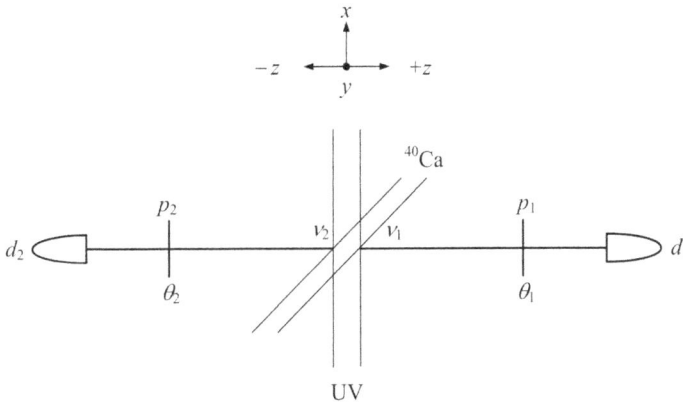

Figure 13.3. Simplified schematics applicable to the Kocher and Commins (1967) optical experiment.

The search to find a measurable effect from local hidden variable theories continued, and in a paper by Freedman and Clauser (1972) an experimental configuration based on the Kocher–Commins experiment was utilized. These authors produced a generalized version of Bell's theorem (Bell 1964) and found that the measured linear polarization correlations agreed with the quantum mechanical predictions, in violation of their new version of Bell's inequalities. Their conclusion was that there was 'strong evidence against local hidden-variable theories' (Freedman and Clauser 1972).

Clauser and Horne (1974) highlighted the incompatibility of a broad class of hidden variable theories, named 'objective local theories', with quantum mechanics and experimental measurements. Of particular interest in this paper is a footnote that refers to the second Wu experiment (Kasday *et al* 1975) as confirming the quantum mechanical predictions; however, the Compton scattering detectors are said to be 'too inefficient' to rule out local theories. On the other hand, Wu and colleagues stated that their $e^+e^- \rightarrow \gamma_1\gamma_2$ measurements 'are thus evidence against local hidden-variable theories' (Kasday *et al* 1975).

13.4 Observations and discussion

Although Wu and colleagues (Kasday *et al* 1975) did refer explicitly to both Bell's inequalities and the Pryce–Ward probability amplitude, this connection was absent in the publications of Clauser *et al* (1969), Freedman and Clauser (1972), and Clauser and Horne (1974). It is also illustrative to point out that the main quantum result exposed by Clauser *et al* (1969) depends on the term $\cos 2\varphi$, where φ is an angular difference. In other words, this appears to be the term $\cos 2(\varphi_1 - \varphi_2)$ derived by Pryce and Ward (1947).

From a historical perspective, none of the publications listed here incorporated the word entanglement in their description. Their main motivating factor appears to have been the interaction between hidden variables local theories and Bell's theorem.

References

Bell J S 1964 On the Einstein–Podolsky–Rosen paradox *Physics* **1** 195–200

Bleuler E and Bradt H L 1948 Correlation between the states of polarization of the two quanta of annihilation radiation *Phys. Rev.* **73** 1398

Bohm D 1951 *Quantum Theory* (Englewood Cliffs, NJ: Prentice-Hall)

Bohm D and Aharonov Y 1957 Discussion of experimental proof for the paradox of Einstein, Rosen, and Podolsky *Phys. Rev.* **108** 1070–6

Clauser J F, Horner M A, Shimony A and Holt R A 1969 Proposed experiment to test local hidden variable theories *Phys. Rev. Lett.* **23** 880–3

Clauser J F and Horner M A 1974 Experimental consequences of objective local theories *Phys. Rev.* D **10** 526–35

Freedman S J and Clauser J F 1972 Experimental test of local hidden-variable theories *Phys. Rev. Lett.* **28** 938–41

Hanna R C 1948 Polarization of annihilation radiation *Nature* **162** 332

Kasday L R, Ullman J D and Wu C S 1975 Angular correlation of Compton-scattered annihilation photons and hidden variables *Il Nuovo Cimento* **25** 633–61

Kocher C A and Commins E D 1967 Polarization correlation of photons emitted in an atomic cascade *Phys. Rev. Lett.* **18** 575–7

Pryce M L H and Ward J C 1947 Angular correlation effects with annihilation radiation *Nature* **160** 435

Wheeler J A 1946 Polyelectrons *Ann. N. Y. Acad. Sci.* **48** 219–38

Wu C S and Shaknov I 1950 The angular correlation of scattered annihilation radiation *Phys. Rev.* **77** 136

IOP Publishing

Fundamentals of Quantum Entanglement

F J Duarte

Chapter 14

The optical quantum entanglement experiments

The Aspect experiments that decisively began as an experimental effort to test for 'local hidden variables' in light of Bell's theorem are described. These experiments utilized an optical version of the original $e^+e^- \rightarrow \gamma_1\gamma_2$ annihilation while using a laser-pumped ^{40}Ca source. These experiments reported strong violation of Bell's inequalities, thus destroying the notion of local hidden variable theories. Still in the early 1980s neither the words quantum entanglement, nor $|\psi\rangle = (|x\rangle_1|y\rangle_2 - |y\rangle_1|x\rangle_2)$, were mentioned.

14.1 Introduction

The annihilation $e^+e^- \rightarrow \gamma_1\gamma_2$ quantum entanglement experiments of Bleuler and Bradt (1948), Hanna (1948), and Wu and Shaknov (1950) were described in chapter 8. As described in chapter 13, the measurements in those experiments were deemed be 'too inefficient' to rule out local hidden variable theories (Clauser and Horner 1974).

Judging from the titles of the papers considered in this section, it is clear that the subject matter motivating this series of experiments was testing hidden variable theories in light of Bell's theorem. This is quite interesting because an even more powerful reason for these experiments would have been to redo the original quantum entanglement experiments (Bleuler and Bradt 1948, Hanna 1948, Wu and Shaknov 1950), previously performed in the x-ray domain, in the visible domain of the electromagnetic spectrum.

14.2 The Aspect experiments

The Aspect experiments are summarized in two papers: the first paper was entitled 'Experimental tests of realistic local theories via Bell's theorem' (Aspect *et al* 1981) and the second paper was entitled 'Experimental realization of Einstein–Podolsky–Rosen–Bohm gedankenexperiment: a new violation of Bell's inequalities' (Aspect *et al* 1982a).

In his first paper on this subject, Aspect does refer to Wu's second experiment (Kasday *et al* 1975) and states that the 'experiments agree with QM predictions. However, because of the lack of efficient polarizers ... strong supplementary assumptions are necessary to interpret these results via Bell's theorem' (Aspect *et al* 1981).

14.2.1 The first Aspect experiment

The first Aspect experiment also utilized ^{40}Ca as the active atomic medium emitting the two quanta in opposite directions. As seen previously, the transitions involved are two cascade transitions in ^{40}Ca: $4p^{21}S_0 - 4p4s^1P_1$ ($\lambda_1 = 551.3$ nm) and $4p4s^1P_1 - 4s^{21}S_0$ ($\lambda_2 = 422.7$ nm) Kocher and Commins (1967). The innovation introduced by Aspect and colleagues was that instead of using a UV arc lamp as the excitation source, they used two lasers to cover the excitation transition $4s^{21}S_0 - 4p^{21}S_0$. This two-photon laser excitation, illustrated in figure 14.1, was accomplished using a single-mode krypton ion laser emitting at $\lambda_{e1} = 406.7$ nm and a tunable dye laser emitting at $\lambda_{e2} = 581$ nm. The two lasers are said to have 'parallel polarizations' (Aspect *et al* 1981). The two lasers were focused on the Ca atomic beam in a counterpropagating configuration perpendicular to the optical axis

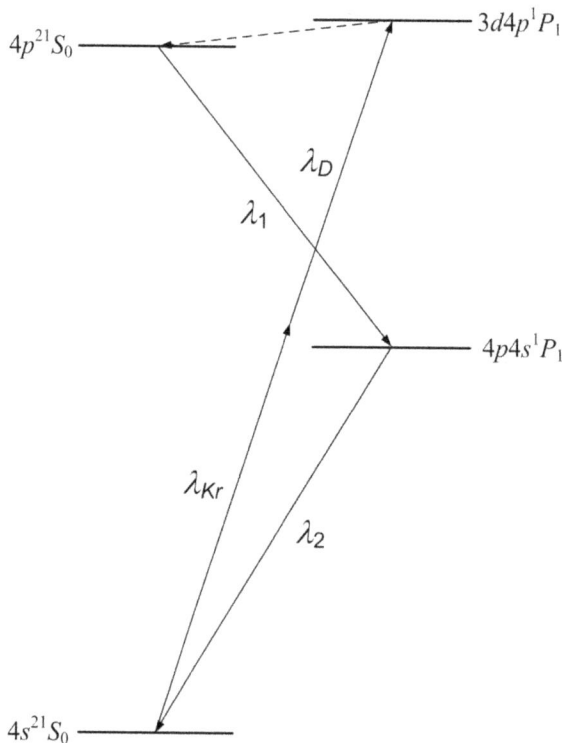

Figure 14.1. Approximate energy level diagram for ^{40}Ca relevant to the Aspect experiments. Aspect *et al* (1981) replaced the UV arc lamp with two counterpropagating lasers at $\lambda_{e1} = 406.7$ nm and $\lambda_{e2} = 581$ nm.

configured by the ^{40}Ca emission. On each side of the emission axis a lens was used to focus the radiation on to a polarizer, followed by a filter, and then a detector. A generic rendition of this experimental configuration is provided in figure 14.2.

Aspect *et al* (1981) utilized the generalized Bell inequalities mentioned previously and as extended by Clauser *et al* (1969) and Clauser and Horner (1974):

$$-1 \leqslant \frac{R(\varphi_1, \varphi_2) - R(\varphi_1, \varphi_2') + R(\varphi_1', \varphi_2') + R(\varphi_1', \varphi_2) - R_1(\varphi_1') - R_2(\varphi_2)}{R_0} \leqslant 0. \quad (14.1)$$

In this inequality $R(\varphi_1, \varphi_2)$, and so on, are coincidence rates with the polarizers oriented at certain angles such as φ_1, φ_2, φ_1', and φ_2'. For instance, the $R(\varphi_1, \varphi_2)$ coincidence rate is measured with the first polarizer in direction φ_1 and the second polarized in direction φ_2, and so on. R_0 is the measured coincidence rate in the absence of polarization analyzers. $R_1(\varphi_1')$ is the measured rate with the second polarizer absent and the first polarized set at φ_1'. $R_2(\varphi_2)$ is the measured rate with the first polarizer absent and the second polarized set at φ_2. The φ_1 and φ_2 angular orientations are illustrated in figure 13.1. Again, the reader will recall that the sum

$$R(\varphi_1, \varphi_2) - R(\varphi_1, \varphi_2') + R(\varphi_1', \varphi_2') + R(\varphi_1', \varphi_2) \quad (14.2)$$

is a slightly modified expression for the left-hand side of Bell's inequality, or

$$|P(x, y) - P(x, y')| + |P(x', y') + P(x', y)| \leqslant 2. \quad (14.3)$$

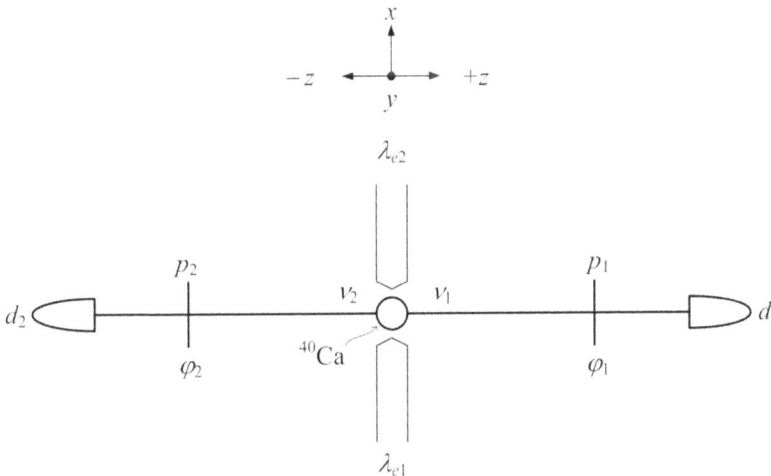

Figure 14.2. Simplified schematics applicable to the Aspect *et al* (1981) experiment. The two counter-propagating excitation lasers at $\lambda_{e1} = 406.7$ nm and $\lambda_{e2} = 581$ nm form an optical axis orthogonal to the axis established by the flow of the ^{40}Ca beam. Again, the green ^{40}Ca transition $4p^2{}^1S_0 - 4p4s^1P_1$ occurs at $\lambda_1 = 551.3$ nm and the blue transition occurs at $\lambda_2 = 422.7$ nm. The angle φ used in the equations is measured at the plane perpendicular to the optical axis and is the difference between φ_1 and φ_2.

The quantum mechanical ratio of coincidence rates predicted by Clauser *et al* (1969) is

$$\frac{R(\varphi)}{R_0} = \varepsilon \, F(\theta) \cos 2\varphi \qquad (14.4)$$

where R_0 was defined above, ε is a factor depending on the efficiency of the polarizers, $F(\theta)$ is a geometrical function, and φ is the angle 'between the polarizer axes' (Clauser *et al* 1969). Here, the reader should remember that in chapter 6 the Pryce–Ward polarization differential angle was defined as $\Delta\varphi = 2(\varphi_1 - \varphi_2)$.

Aspect and colleagues report a 'perfect agreement with quantum mechanics' and a strong violation of their applicable Bell inequality (Aspect *et al* 1981). An interesting result from these experiments is that the strongest violations of the generalized Bell's inequalities were found when the polarizers were set at interval angular segments of 22.5° (Aspect *et al* 1981) as predicted by Clauser *et al* (1969).

14.2.2 The second Aspect experiment

The second Aspect experiment replaced the straightforward polarizer analyzer by a polarizing beam splitter in conjunction with two detectors (Aspect *et al* 1982a), as illustrated in figure 14.3. This type of detection arrangement has become ubiquitous in current quantum entanglement experiments.

Again using Ca as the emission medium and the same excitation configuration as in the first experiment, Aspect and colleagues again relied on the $4p^{21}S_0 - 4p4s^1P_1$ ($\lambda_1 = 551.3$ nm) and $4p4s^1P_1 - 4s^{21}S_0$ ($\lambda_2 = 422.7$ nm) transitions of ^{40}Ca to perform polarization correlation measurements. These authors reported 'the strongest violation of Bell's inequalities ever achieved, and excellent agreement with quantum mechanics' (Aspect *et al* 1982a).

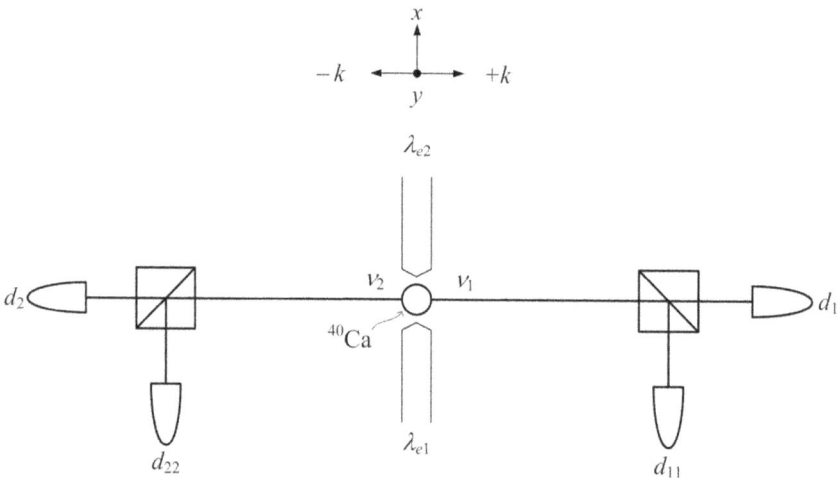

Figure 14.3. Simplified schematics applicable to the second Aspect experiment. In this case the polarizer analyzers p_1 and p_2 are replaced by polarizing beam splitters.

14.2.3 The third Aspect experiment

Up to now all the experiments considered here have involved static conditions in which the polarizing analyzers are kept stationary in their angular settings while experimental runs are conducted. These static experiments do not satisfy the conditions of the gedankenexperiment proposed by Bohm and Aharonov (1957) in which the polarization settings are changed during the propagation of the quanta.

An experiment including time-varying polarization analyzers was reported by Aspect *et al* (1982b). In this experiment, each detection station was composed of an electro–optical Bragg grating, which acts as a time-dependent beam splitter, followed by two polarizers: one polarizer along the original optical axis and a second polarizer along the path of the diffracted beam. The Bragg grating operates at a frequency of ~ 50 MHz (Aspect *et al* 1982b). In this experimental configuration the predetermined angles φ_1 and φ_1' are at one measuring station with φ_1 at the polarizer analyzer deployed orthogonal to the main optical axis and φ_1' at the polarizer analyzer deployed orthogonal to the diffracted beam. The angles φ_2 and φ_2' are at the complementary measuring station with φ_2 at the polarizer analyzer deployed orthogonal to the main optical axis and φ_2' at the polarizer analyzer deployed orthogonal to the diffracted beam. See figure 14.4.

Again, Aspect *et al* (1982b) utilized the generalized Bell inequalities mentioned previously and as extended by Clauser *et al* (1969) and Clauser and Horner (1974):

$$
\begin{aligned}
-1 \leqslant &\frac{R(\varphi_1, \varphi_2)}{R(a_1, a_2)} - \frac{R(\varphi_1, \varphi_2')}{R(a_1, a_2')} + \frac{R(\varphi_1', \varphi_2')}{R(a_1', a_2')} \\
+ &\frac{R(\varphi_1', \varphi_2)}{R(a_1', a_2)} - \frac{R(\varphi_1', a_2)}{R(a_1', a_2)} - \frac{R(a_1, \varphi_2)}{R(a_1, a_2)} \leqslant 0.
\end{aligned}
\tag{14.5}
$$

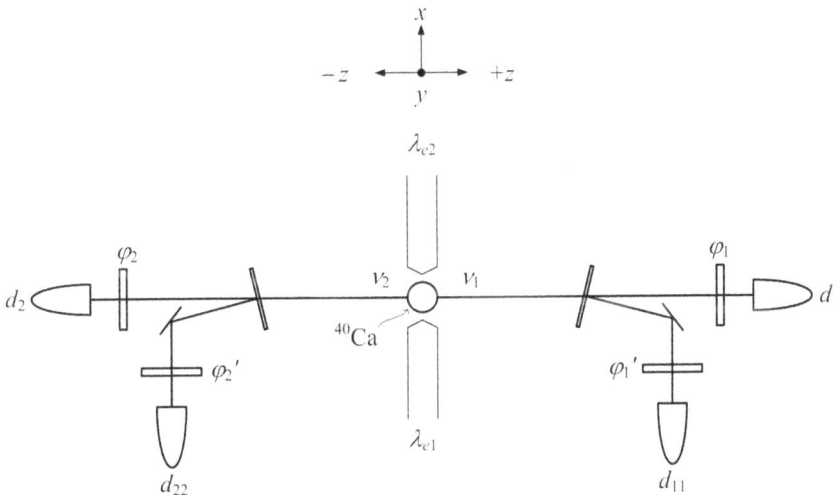

Figure 14.4. Schematics applicable to the third Aspect experiment describing the positions of φ_1, φ_1', φ_2, and φ_2' relative to the optical axes.

In this inequality, $R(\varphi_1, \varphi_2)$, $R(\varphi_1, \varphi_2')$, $R(\varphi_1', \varphi_2')$, and $R(\varphi_1', \varphi_2)$ are single run coincidence rates. $R(a_1, a_2)$, $R(a_1, a_2')$, $R(a_1', a_2')$, and $R(a_1', a_2)$ are said to be corresponding coincidence rates with all polarizers absent (a) from the configuration.

$R(\varphi_1', a_2)$ and $R(a_1, \varphi_2)$ are measured coincidence rates with the indicated polarizers removed from each end of the optical axis (Aspect $et\ al$ 1982b). The reported result obtained for

$$(\varphi_1, \varphi_2) = (\varphi_1', \varphi_2) = (\varphi_1', \varphi_2') = \frac{\pi}{8}$$

and

$$(\varphi_1, \varphi_2') = \frac{3\pi}{8}$$

is

$$\frac{R(\varphi_1, \varphi_2)}{R(a_1, a_2)} - \frac{R(\varphi_1, \varphi_2')}{R(a_1, a_2')} + \frac{R(\varphi_1', \varphi_2')}{R(a_1', a_2')}$$
$$+ \frac{R(\varphi_1', \varphi_2)}{R(a_1', a_2)} - \frac{R(\varphi_1', a_2)}{R(a_1', a_2)} - \frac{R(a_1, \varphi_2)}{R(a_1, a_2)} = 0.101 \pm 0.020$$

which represents a violation of the inequality by five standards of deviation. The applicable quantum mechanical coincidence ratio is quoted as 0.112 (Aspect $et\ al$ 1982b).

This third Aspect experiment might be regarded as one the first attempts to close a so-called loop hole in this class of experiments. Here, in reference to Bell (1964), Aspect and colleagues write, 'the locality condition would then become a consequence of Einstein's causality, preventing any faster than light influence' (Aspect $et\ al$ 1982b).

14.3 Observations and discussion

Although Wu and colleagues (Kasday $et\ al$ 1975) did refer explicitly to both Bell's inequalities and the Pryce–Ward probability amplitude, this connection was absent in the publications of Clauser $et\ al$ (1969), Freedman and Clauser (1972), Clauser and Horner (1974), Aspect $et\ al$ (1981), and Aspect $et\ al$ (1982a, 1982b). It is also illustrative to point out that the main quantum result exposed by Clauser $et\ al$ (1969) depends on the term $\cos 2\varphi$, where φ stands for an angular difference, which should be compared with the term $\cos 2(\varphi_1 - \varphi_2)$ derived by Pryce and Ward (1947).

From a historical perspective, none of the publications listed here incorporated the word entanglement in their description. Their main motivating factor appears to have been the interaction between hidden variables local theories and Bell's theorem. However, by the second Aspect experiment the emphasis was clearly on the agreement of experimental results with quantum mechanics and the strong violation of Bell's inequalities (Aspect $et\ al$ 1982a, 1982b).

One observation due here is that in the Aspect experiments there is a deviation from one of the initial requirements suggested by Wheeler (1946) and utilized by Ward (1949) in his derivation of the probability amplitude $|\psi\rangle = (|x_1, y_2\rangle - |y_1, x_2\rangle)$. That requirement is associated with *indistinguishability* and the idea that $|p_1| = |p_2|$, which implies that $\lambda_1 = \lambda_2$, but in Aspect's experiments $\lambda_1 = 551.3$ and $\lambda_2 = 422.7$ nm.

References

Aspect A, Grangier P and Roger G 1981 Experimental tests of realistic local theories via Bell's theorem *Phys. Rev. Lett.* **47** 460–3

Aspect A, Grangier P and Roger G 1982a Experimental realization of Einstein–Podolsky–Rosen–Bohm gedanken experiment: a new violation of Bell's inequalities *Phys. Rev. Lett.* **49** 91–4

Aspect A, Grangier P and Roger G 1982b Experimental test of Bell's inequality using time-varying analyzers *Phys. Rev. Lett.* **49** 1804–7

Bell J S 1964 On the Einstein–Podolsky–Rosen paradox *Physics* **1** 195–200

Bleuler E and Bradt H L 1948 Correlation between the states of polarization of the two quanta of annihilation radiation *Phys. Rev.* **73** 1398

Bohm D and Aharonov Y 1957 Discussion of experimental proof for the paradox of Einstein, Rosen, and Podolsky *Phys. Rev.* **108** 1070–6

Clauser J F, Horner M A, Shimony A and Holt R A 1969 Proposed experiment to test local hidden variable theories *Phys. Rev. Lett.* **23** 880–3

Clauser J F and Horner M A 1974 Experimental consequences of objective local theories *Phys. Rev.* D **10** 526–35

Freedman S J and Clauser J F 1972 Experimental test of local hidden-variable theories *Phys. Rev. Lett.* **28** 938–41

Hanna R C 1948 Polarization of annihilation radiation *Nature* **162** 332

Kasday L R, Ullman J D and Wu C S 1975 Angular correlation of Compton-scattered annihilation photons and hidden variables *Il Nuovo Cimento* **25** 633–61

Kocher C A and Commins E D 1967 Polarization correlation of photons emitted in an atomic cascade *Phys. Rev. Lett.* **18** 575

Pryce M L H and Ward J C 1947 Angular correlation effects with annihilation radiation *Nature* **160** 435

Ward J C 1949 *Some Properties of the Elementary Particles* (Oxford: Oxford University Press)

Wheeler J A 1946 Polyelectrons *Ann. N. Y. Acad. Sci.* **48** 219–38

Wu C S and Shaknov I 1950 The angular correlation of scattered annihilation radiation *Phys. Rev.* **77** 136

Chapter 15

The quantum entanglement probability amplitude 1947–1992

The Pryce–Ward probability amplitude, $|\psi\rangle = (|x\rangle_1|y\rangle_2 - |y\rangle_1|x\rangle_2)$, in its various formats as it appeared in the 1947–92 literature is presented.

15.1 Introduction

If there is one lesson to be learnt from this monograph is that quantum entanglement is completely explained by the probability amplitude

$$|\psi\rangle = \frac{1}{\sqrt{2}}(|x\rangle_1|y\rangle_2 - |y\rangle_1|x\rangle_2) \tag{15.1}$$

whose physics was explained in chapter 7. Here, a brief summary on the emergence of this iconic probability amplitude is given.

15.2 The quantum entanglement probability amplitude 1947–92

Here, the timeline of the appearance of equations of the form of equation (15.1) is presented by year of publication.

15.2.1 1947–9

The first time that the quantum entanglement probability amplitude was set on paper was done, using Dirac's notation (Dirac 1939, 1978) by Pryce and Ward (1947), as an intermediate step in the derivation of the correct quantum cross-section for scattering measurements applicable to the polarization quantum entanglement of two indistinguishable quanta propagating in opposite directions (see chapter 6). In this regard, the probability amplitudes prior to normalization can be written as

$$|\psi\rangle = (|x, y\rangle - |y, x\rangle) \tag{15.2}$$

doi:10.1088/2053-2563/ab2b33ch15

which, using Dirac's identities, is equivalent to (Duarte 2014)

$$|\psi\rangle = (|x\rangle_1|y\rangle_2 - |y\rangle_1|x\rangle_2) \tag{15.3}$$

and once the usual normalization is performed can be immediately expressed as equation (15.1). In other words, the physics of equations (15.1)–(15.3) is completely equivalent. Ward presented his derivation, discussed in chapter 6, in his doctoral thesis (Ward 1949).

15.2.2 1948

Using a wave function format, rather than Dirac's notation, and in the absence of derivation, Snyder *et al* (1948) wrote an expression of the form

$$\psi = \frac{1}{\sqrt{2}}(\psi_a(1)\psi_b(2) + \psi_c(1)\psi_d(2)). \tag{15.4}$$

15.2.3 1951

Using a wave function format, Bohm (1951) wrote

$$\psi_1 = \frac{1}{\sqrt{2}}(\psi_+(1)\psi_-(2) + \psi_-(1)\psi_+(2)) \tag{15.5}$$

In his book explanation, Bohm (1951) did not cite Snyder *et al* (1948), nor did he mention the Dirac notation alternative.

15.2.4 1957

Using a wave function format, and in the absence of derivation in a paper that cited Snyder *et al* (1948), Bohm and Aharonov (1957) wrote

$$\psi = \frac{1}{\sqrt{2}}(\psi_+(1)\psi_-(2) - \psi_-(1)\psi_+(2)). \tag{15.6}$$

15.2.5 1975

Using Dirac's *bra–ket* notation (Dirac 1939, 1978), while citing Pryce and Ward (1947), Wu and colleagues (Kasday *et al* 1975) wrote

$$\psi = \frac{1}{\sqrt{2}}(XY - YX). \tag{15.7}$$

15.2.6 1990

Using Dirac's *bra–ket* notation, in the absence of derivation and without literature origin, and in reference to spins, Greenberger *et al* (1990) wrote

$$|\psi\rangle = \frac{1}{\sqrt{2}}(|+\rangle_1|-\rangle_2 - |-\rangle_1|+\rangle_2). \tag{15.8}$$

15.2.7 1992

Using Dirac's *bra–ket* notation, in the absence of derivation and without literature origin, and in reference to spins, Bennett and Wiesner (1992) wrote

$$|\psi\rangle = \frac{1}{\sqrt{2}}(|\uparrow\downarrow\rangle - |\downarrow\uparrow\rangle). \tag{15.9}$$

15.3 Observations and discussion

In the mid to late 1990s the probability amplitude represented by equation (15.1), in its various formats, became a widely used equation in the armamentarium of quantum entanglement physicists. Undoubtedly, it has become a true icon. Its origin, however, is almost never discussed (Duarte 2012). It was, and largely still is, as if everyone knew where it came from and its meaning assumed to be obvious. Perhaps it might be conjectured that the 'similarity' of the right-hand side of equation (15.1) with other iconic quantum equations such as (Born *et al* 1926)

$$(PQ - QP) = \frac{h}{2\pi i} \tag{15.10}$$

dampened the interest of the quantum entanglement community to find out the origin of equation (15.1). However, that is a tenuous analogy at best and there is no published evidence to support this conjecture.

Furthermore, as has been seen in previous chapters of this book,

$$|\psi\rangle = \frac{1}{\sqrt{2}}(|x\rangle_1|y\rangle_2 - |y\rangle_1|x\rangle_2)$$

was not even mentioned in the mainstream literature of the field in the crucial 1957–82 period when the main attention was focused on EPR arguments, hidden variable theories, and Bell's theorem.

This will go down as one of the strangest episodes in the history of physics where the crucial and essential equation to the most important branch of quantum optics vanished for a period and then re-emerged out of thin air more than 40 years later … meanwhile, the inventors, or discoverers, of the equation were apparently unknown to the users and beneficiaries of the equation. Indeed, an ineffable omission.

References

Bennet C H and Wiesner S J 1992 Communication via one- and two-particle operators on Einstein–Podolsky–Rosen states *Phys. Rev. Lett.* **69** 2881–4

Bohm D 1951 *Quantum Theory* (Englewood Cliffs, NJ: Prentice Hall)

Bohm D and Aharonov Y 1957 Discussion of experimental proof for the paradox of Einstein, Rosen, and Podolsky *Phys. Rev.* **108** 1070–6

Born M, Heisenberg W and Jordan P 1926 Zur quantenmechanik II *Z. Phys.* **35** 557–615

Dirac P A M 1939 A new notation for quantum mechanics *Math. Proc. Camb. Philos. Soc.* **35** 416–8

Dirac P A M 1978 *The Principles of Quantum Mechanics* 4th edn (Oxford: Oxford University Press)

Duarte F J 2012 The origin of quantum entanglement experiments based on polarization measurements *Eur. Phys. J.* H **37** 311–8

Duarte F J 2014 *Quantum Optics for Engineers* (New York: CRC)

Greenberger D M, Horne M A, Shimony A and Zeilinger A 1990 Bell's theorem without inequalities *Am. J. Phys.* **58** 1131–43

Kasday L R, Ullman J D and Wu C S 1975 Angular correlation of Compton-scattered annihilation photons and hidden variables *Il Nuovo Cimento* **25** 633–61

Pryce M L H and Ward J C 1947 Angular correlation effects with annihilation radiation *Nature* **160** 435

Snyder H S, Pasternack S and Hornbostel J 1948 Angular correlations of scattered annihilated radiation *Phys. Rev.* **73** 440–8

Ward J C 1949 *Some Properties of the Elementary Particles* (Oxford: Oxford University Press)

IOP Publishing

Fundamentals of Quantum Entanglement

F J Duarte

Chapter 16

The GHZ probability amplitudes

The GHZ probability amplitudes for three and four quanta are introduced.

16.1 Introduction

The Pryce–Ward probability amplitude (Pryce and Ward 1947, Ward 1949)

$$|\psi\rangle = \frac{1}{\sqrt{2}}(|x\rangle_1 |y\rangle_2 - |y\rangle_1 |x\rangle_2) \tag{16.1}$$

applies to two indistinguishable quanta, γ_1 and γ_2, propagating along a common optical axis in opposite directions. That is, there are two particles involved ($n = 2$), two propagation channels ($N = 2$), $+z$ and $-z$, and two polarization alternatives $|x\rangle$ and $|y\rangle$ that are orthogonal to each other. In short, the Pryce–Ward probability amplitude applies to the case of $n = N = 2$.

Greenberger *et al* (1990) reintroduced the Pryce–Ward probability amplitude presented in a slightly different notation,

$$|\psi\rangle = \frac{1}{\sqrt{2}}(|+\rangle_1 |-\rangle_2 - |-\rangle_1 |+\rangle_2) \tag{16.2}$$

in a framework of spin-1/2 particles. Here, $|+\rangle_1$ represents the spin-up of particle 1 (equivalent to $|x\rangle_1$, or quanta 1 being polarized in the $|x\rangle$ state), $|-\rangle_2$ represents the spin-down of particle 2 (equivalent to $|y\rangle_2$, or quanta 2 being polarized in the $|y\rangle$ state), and so on for $|-\rangle_1$ and $|+\rangle_2$. Greenberger *et al* (1990) did not explain the origin of equation (16.2), nor did they derive it from first principles. This is not meant as a criticism to these authors, it is simply stated to indicate that by the early 1990s equation (16.1) had probably already reached folklore status and everyone knew that it was out there.

doi:10.1088/2053-2563/ab2b33ch16

16.2 The GHZ probability amplitudes

A probability amplitude beyond the fundamental case of $n = N = 2$ was introduced by Greenberger *et al* (1990). This probability amplitude is known in the literature as the GHZ probability amplitude after the initials of the three original co-authors (Greenberger, Horne, and Zeilinger).

The GHZ probability amplitude deals with four propagation channels ($N = 4$) and four particles ($n = 4$), as illustrated in figure 16.1. The GHZ equation was originally developed for spin-1/2 particles and has the following form (Greenberger *et al* (1990)):

$$|\psi\rangle = \frac{1}{\sqrt{2}}(|+\rangle_1 |+\rangle_2 |-\rangle_3 |-\rangle_4 - |-\rangle_1 |-\rangle_2 |+\rangle_3 |+\rangle_4). \qquad (16.3)$$

The reader should be able to discern that the first term, within the parenthesis, refers to particles 1 and 2 with spins up and particles 3 and 4 with spins down. Along the same lines, the second term in this equation refers to particles 1 and 2 with spins down and particles 3 and 4 with spins up. In summary, this probability amplitude applies to the case of $n = N = 4$ (see figure 16.1).

For the case of three particles ($n = N = 3$), outside the realm of spins, Greenberger *et al* (1990) introduced, in the absence of a derivation,

$$|\psi\rangle = \frac{1}{\sqrt{2}}(|a\rangle_1 |b\rangle_2 |c\rangle_3 + |a'\rangle_1 |b'\rangle_2 |c'\rangle_3). \qquad (16.4)$$

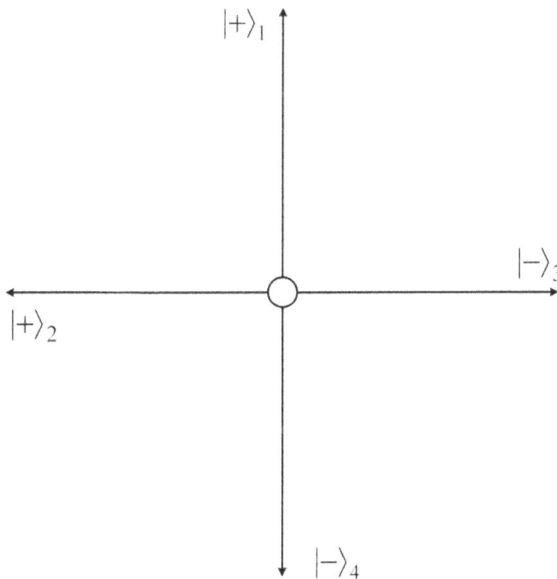

Figure 16.1. Simplified $n = N = 4$ quantum entanglement configuration applicable to the scheme introduced by Greenberger *et al* (1990). With spin-1/2 particles the spin-up (+) and spin-down (−) alternatives are separated using a Stern–Gerlach apparatus. For a description of the Stern–Gerlach physics, for spin-1/2 and spin-1 particles, the reader should refer to Feynman *et al* (1965).

The derivation of equation (16.3) is described by Greenberger *et al* (1990) and is not considered here due to the fact that it contains assumptions and concepts that are not within the scope of this book. Two observations are relevant here: for $n = N = 4$ one would have expected four terms in equation (16.3) and $1/\sqrt{4}$; for $n = N = 3$ one would have expected three terms in equation (16.4) and $1/\sqrt{3}$. The case for probability amplitudes applicable to $n = N = 2^1, 2^2, 2^3 \ldots 2^n$ is considered in chapters 17 and 18 from the ground up in a consistent and systematic approach. The special case for $n = N = 3, 6$ is considered in detail in chapter 19.

16.3 Observations and discussion

The significance of the work by Greenberger *et al* (1990) was three-fold:

1. They reintroduced the original probability amplitude for quantum entanglement, albeit in a slightly modified notation, squarely onto the platform of the philosophical path, the EPR argument (Einstein *et al* 1935), and Bell's theorem (Bell 1964).
2. They proved the rotational invariance of the probability amplitude for quantum entanglement for the case of $n = N = 2$ (equation (16.1) or (16.2)).
3. They were able to recast Bell's theorem without the need of inequalities.
4. Moreover, they found that the EPR program 'contradicts quantum mechanics even for the case of perfect correlations'. These authors go on to claim that for $n = N = 3$ and $n = N = 4$ there is an even stronger incompatibility than that observed for $n = N = 2$ (Greenberger *et al* 1990).

References

Bell J S 1964 On the Einstein–Podolsky–Rosen paradox *Physics* **1** 195–200

Einstein A, Podolsky B and Rosen N 1935 Can quantum mechanical description of physical reality be considered complete? *Phys. Rev.* **47** 777–80

Feynman R P, Leighton R B and Sands M 1965 *The Feynman Lectures on Physics* vol III (Reading, MA: Addison-Wesley)

Greenberger D M, Horne M A, Shimony and Zeilinger A 1990 Bell's theorem without inequalities *Am. J. Phys.* **58** 1131–43

Pryce M L H and Ward J C 1947 Angular correlation effects with annihilation radiation *Nature* **160** 435

Ward J C 1949 *Some Properties of the Elementary Particles* (Oxford: Oxford University Press)

Chapter 17

The interferometric derivation of the quantum entanglement probability amplitude for $n = N = 2$

The interferometric derivation of the Pryce–Ward probability amplitude, $|\psi\rangle = (|x\rangle_1 |y\rangle_2 - |y\rangle_1 |x\rangle_2)$, applicable to two entangled quanta ($n = 2$) and two propagation channels ($N = 2$), that is $n = N = 2$, is described in detail using the Dirac–Feynman principle and Dirac's identities.

17.1 Introduction

In chapter 6 the heuristic derivation of the probability amplitude for two quanta propagating in opposite directions, with entanglement polarizations (Pryce and Ward 1947, Ward 1949)

$$|\psi\rangle = \frac{1}{\sqrt{2}}(|x\rangle_1 |y\rangle_2 - |y\rangle_1 |x\rangle_2) \tag{17.1}$$

was outlined. Chapter 11 described the physics related to the Hamiltonian for a two-state system, as described by Feynman *et al* (1965), leading to equations of the form

$$|\psi\rangle = \frac{1}{\sqrt{2}}(|B\rangle - |A\rangle) \tag{17.2}$$

which are a preamble to the arrival to equation (17.1).

In this chapter the probability amplitude for quantum entanglement is derived from first principles utilizing the Dirac–Feynman principle (Dirac 1939, 1978, Feynman *et al* 1965)

$$\langle x|s\rangle = \sum_{j=1}^{N} \langle x|j\rangle \langle j|s\rangle \tag{17.3}$$

doi:10.1088/2053-2563/ab2b33ch17

as introduced in chapter 2.

The case considered in this chapter applies to two quanta ($n = 2$) and two propagation channels ($N = 2$) and is thus referred to as $n = N = 2$.

17.2 The meaning of the Dirac–Feynman probability amplitude

Although it could have been done in chapter 2, it is now advisable to think about the meaning of the Dirac–Feynman probability amplitude

$$\langle x|s \rangle = \sum_{j=1}^{N} \langle x|j \rangle \langle j|s \rangle.$$

This equation is written yet once more since its importance is utterly crucial to the field of quantum optics. Up to now this equation has been extensively applied in a nameless fashion. In this monograph it is called the Dirac–Feynman probability amplitude because it was Feynman who presented and championed it in his famous lectures (Feynman *et al* 1965). At the same time, Feynman's presentation was heavily influenced by Dirac's *bra–ket* notation and Dirac's principles as outlined in his book (Dirac 1978).

The meaning of this equation is quite succinct and yet profound:

1. It says that, 'all the indistinguishable photons illuminate the array of N slits, or grating, simultaneously. If only one photon propagates ... then that individual photon illuminates the whole array of N slits simultaneously' (Duarte 2003).
2. It also says that in order to obtain the correct result, the interaction of the probability amplitude via every slit, under illumination, with the probability amplitude via every other slit in the array must be included in the calculation.

For example, if an array of 2000 slits is under illumination then the interaction of the probability amplitude of slit number 1 with all the other 1999 slits must be accounted for, and so on. In other words, *all the probability amplitudes are interconnected.*

17.3 The derivation of the quantum entanglement probability amplitude

The experimental diagram applicable to the fundamental configuration of quantum entanglement is depicted in figure 17.1. Here, s is a photon source of ν_1 and ν_2 emitted in the $+z$ and $-z$ directions, respectively. The quanta propagate along the optical axis, in the $+z$ and $-z$ directions, toward the polarization analyzers p_1 and p_2 and the detectors d_1 and d_2, respectively.

The first essential, and experimentally sound, assumption is that the quanta emitted by the source are *indistinguishable*, that is, $\nu_1 = \nu_2 = \nu$. Then, applying the Dirac interferometric principle stated in equation (17.3),

$$\langle d|s \rangle = \langle d|p_2 \rangle \langle p_2|s \rangle + \langle d|p_1 \rangle \langle p_1|s \rangle. \tag{17.4}$$

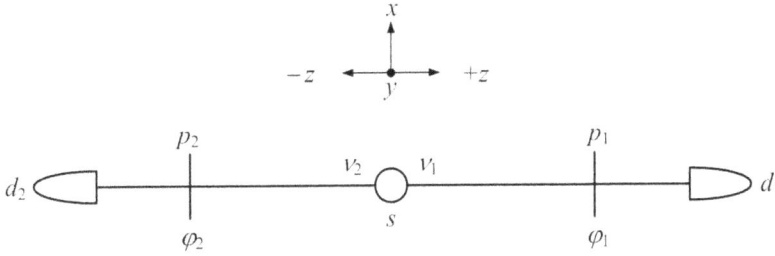

Figure 17.1. Experimental diagram applicable to quantum entanglement. Here, s is the photon source of v_1 and v_2 emitted in the $+z$ and $-z$ directions, respectively. The two linear polarization axes are indicated as x and y, which are perpendicular to each other. The angles φ_1 and φ_2 are the polarization angles measured on a plane perpendicular to the propagation axis while d_1 and d_2 are the corresponding detectors.

Implicitly, this equation assumes that $d_1 = d_2 = d$, which is quite reasonable for a pair of matched detectors. Abstracting d from equation (17.4), this expression can be restated as

$$|s\rangle = |p_2\rangle\langle p_2|s\rangle + |p_1\rangle\langle p_1|s\rangle. \qquad (17.5)$$

Next, using the Dirac identity (see chapter 2)

$$|\psi\rangle = |j\rangle\langle j|\psi\rangle \qquad (17.6)$$

allows the expression of

$$|D\rangle_2 = |p_2\rangle\langle p_2|s\rangle \qquad (17.7)$$

and

$$|D\rangle_1 = |p_1\rangle\langle p_1|s\rangle. \qquad (17.8)$$

Substituting identities (17.7) and (17.8) into equation (17.5) yields an equation of the form

$$|s\rangle = (|D\rangle_2 + |D\rangle_1). \qquad (17.9)$$

Equation (17.9) is a probability amplitude $|s\rangle$ representing the linear combination of two probability amplitudes, $|D\rangle_1$ and $|D\rangle_2$, and applies equally to an interferometric situation involving a single photon, or a population of indistinguishable photons, and to the quantum entanglement framework being developed here. The crucial difference occurs at this very stage that invokes the use of the Dirac identity for 'similar particles' (see chapter 2):

$$|X\rangle = |a\rangle_1 |b\rangle_2 |c\rangle_3...|g\rangle_n. \qquad (17.10)$$

This very step diverges from a purely interferometric situation given that the possibility of different quanta, quanta 1 and quanta 2, is allowed with each quanta in different alternative states. That is, state $|x\rangle$ and state $|y\rangle$. Rewriting $|D\rangle_1$ and $|D\rangle_2$ in terms of polarization states as per identity (17.10), and using $|C\rangle_1$ and $|C\rangle_2$ to distinguish this from a pure interferometric event, leads to

$$|s\rangle = (|C\rangle_2 + |C\rangle_1) \tag{17.11}$$

where

$$|C\rangle_1 = |x\rangle_1\, |y\rangle_2 \tag{17.12}$$

and

$$|C\rangle_2 = |y\rangle_1\, |x\rangle_2. \tag{17.13}$$

Inserting identities (17.12) and (17.13) in the expression for the probability amplitude in equation (17.11) leads directly to (Duarte 2013a, 2013b, 2014)

$$|s\rangle = (|x\rangle_1\, |y\rangle_2 + |y\rangle_1\, |x\rangle_2). \tag{17.14}$$

Following normalization (see chapter 11), and designating $|s\rangle$ as $|\psi\rangle$, equation (17.14) can be expressed as

$$|\psi\rangle_+ = \frac{1}{\sqrt{2}}(|x\rangle_1\, |y\rangle_2 + |y\rangle_1\, |x\rangle_2) \tag{17.15}$$

and its linear combination is

$$|\psi\rangle_- = \frac{1}{\sqrt{2}}(|x\rangle_1\, |y\rangle_2 - |y\rangle_1\, |x\rangle_2) \tag{17.16}$$

which is the iconic probability amplitude for quantum entanglement.

For notational functionality, equations (17.15) and (17.16) can be expressed in a 2×2 arrangement as

$$\begin{array}{ll} +|C\rangle_1 & +|C\rangle_2 \\ +|C\rangle_1 & -|C\rangle_2. \end{array} \tag{17.17}$$

17.4 Identical states of polarization

Going back to equation (17.11), if the two quanta pairs have identical states of polarization, then equations (17.12) and (17.13) become

$$|C\rangle_1 = |x\rangle_1\, |x\rangle_2 \tag{17.18}$$

and

$$|C\rangle_2 = |y\rangle_1\, |y\rangle_2 \tag{17.19}$$

so that

$$|\psi\rangle^+ = \frac{1}{\sqrt{2}}(|x\rangle_1\, |x\rangle_2 + |y\rangle_1\, |y\rangle_2) \tag{17.20}$$

and

$$|\psi\rangle^{-} = \frac{1}{\sqrt{2}}(|x\rangle_1\,|x\rangle_2 - |y\rangle_1\,|y\rangle_2). \tag{17.21}$$

These types of probability amplitudes, $|\psi\rangle^{+}$ and $|\psi\rangle^{-}$, are of interest in quantum computing. In summary, the whole family of probability amplitudes relevant to two quanta is represented by $|\psi\rangle_{+}$, $|\psi\rangle_{-}$, $|\psi\rangle^{+}$, and $|\psi\rangle^{-}$.

17.5 Discussion

The probability amplitude for quantum entanglement was first put down on paper in 1947 by Pryce and Ward, and its semi-heuristic derivation was made public in 1949 by Ward (Pryce and Ward 1947, Ward 1949). Since Ward's effort until 2013, there has been no record in the open literature of additional derivations. As already documented in previous chapters, all of the research related to quantum entanglement in the post-Bohm–Aharonov epoch (Bohm and Aharonov 1957) centered around the EPR arguments (Einstein *et al* 1935) and Bell's theorem (Bell 1964). As explained in chapter 11, in 1965, Feynman wrote down the physics, via a two-level Hamiltonian, that led to equation (17.2); however, he stopped there and did not consider quantum entanglement (Feynman *et al* 1965).

The derivation detailed here shows that the Dirac–Feynman interferometric principle

$$\langle x|s\rangle = \sum_{j=1}^{N}\langle x|j\rangle\,\langle j|s\rangle$$

central to the development of generalized interferometric probability equations (see chapter 2) is also a foundation of quantum entanglement. What's important to highlight is that the derivation of the quantum entanglement probability amplitude, *à la Dirac*, flows naturally; it is transparent and straightforward.

References

Bell J S 1964 On the Einstein–Podolsky–Rosen paradox *Physics* **1** 195–200
Bohm D and Aharanov Y 1957 Discussion of experimental proof for the paradox of Einstein, Rosen, and Podolsky *Phys. Rev.* **108** 1070–6
Dirac P A M 1939 A new notation for quantum mechanics *Math. Proc. Camb. Philos. Soc.* **35** 416–8
Dirac P A M 1978 *The Principles of Quantum Mechanics* 4th edn (Oxford: Oxford University Press)
Duarte F J 2003 *Tunable Laser Optics* (New York: Elsevier)
Duarte F J 2013a The probability amplitude for entangled polarizations: an interferometric approach *J. Mod. Opt.* **60** 1585–7
Duarte F J 2013b Tunable laser optics: applications to optics and quantum optics *Prog. Quantum Electron.* **37** 326–47
Duarte F J 2014 *Quantum Optics for Engineers* (New York: CRC)

Einstein A, Podolsky B and Rosen N 1935 Can quantum mechanical description of physical reality be considered complete? *Phys. Rev.* **47** 777–80

Feynman R P, Leighton R B and Sands M 1965 *The Feynman Lectures on Physics* vol III (Reading, MA: Addison-Wesley)

Pryce M L H and Ward J C 1947 Angular correlation effects with annihilation radiation *Nature* **160** 435

Ward J C 1949 *Some Properties of the Elementary Particles* (Oxford: Oxford University Press)

Chapter 18

The interferometric derivation of the quantum entanglement probability amplitude for $n = N = 2^1, 2^2, 2^3, 2^4, \ldots 2^r$

The interferometric derivation of probability amplitudes for $n = N = 2^1, 2^2, 2^3, 2^4, \ldots 2^r$, where $r = 1, 2, 3, 4, 5, \ldots$, is described in detail using the Dirac–Feynman principle and Dirac's identities. Here, n is the number of quanta and N is the number of propagation channels.

18.1 Introduction

Chapter 17 outlined the interferometric derivation, *à la Dirac*, of the probability amplitude for quantum entanglement applicable to two quanta ($n = 2$) and two propagation channels ($N = 2$)

$$|\psi\rangle = \frac{1}{\sqrt{2}}(|x\rangle_1 |y\rangle_2 - |y\rangle_1 |x\rangle_2). \tag{18.1}$$

This case is the most fundamental of the entanglement alternatives and is referred to as $n = N = 2$. In this chapter, the same derivational approach utilized in chapter 17 is applied to the cases of $n = N = 4$, $n = N = 8$, and $n = N = 16$.

Furthermore, generalized equations applicable to $n = N = 2^1, 2^2, 2^3, 2^4, \ldots 2^r$ or simply $n = N = 2^r$, where $r = 1, 2, 3, 4, 5, \ldots$, are also given.

The material presented in this chapter is a summary and extension of various publications (Duarte 2013a, 2013b, 2014, 2015, 2016, 2018, Duarte and Taylor 2017).

18.2 The quantum entanglement probability amplitude for $n = N = 4$

The experimental diagram applicable to the $n = N = 4$ configuration for quantum entanglement is depicted in figure 18.1. Here, s is the photon source of two pairs of

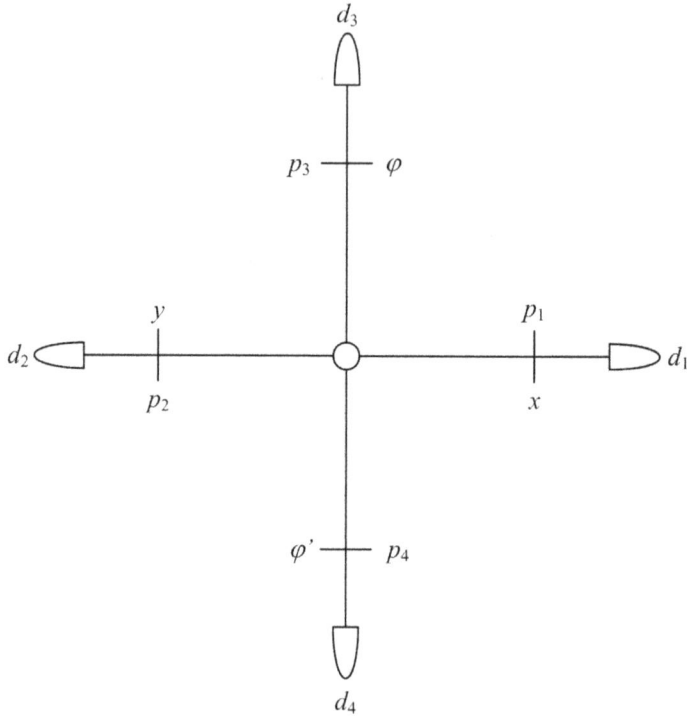

Figure 18.1. Experimental diagram applicable to quantum entanglement for $n = N = 2^2$.

quanta emitted in opposite directions relative to each other: (ν_1, ν_2) and (ν_3, ν_4). In terms of conventional axis labeling, (ν_1, ν_2) are emitted in the $+z$ and $-z$ directions while (ν_3, ν_4) are emitted in the $+x$ and $-x$ directions, respectively. In order to simplify the notation and axes labeling, in reference to figure 18.1, traditional axes labels are obviated and only polarization alternatives are assigned in pairs; in this case, (x, y) along the $+z$ and $-z$ directions, and (φ, φ') along the corresponding orthogonal direction.

Again, the first essential and experimentally sound assumption is that the quanta emitted by the source are *indistinguishable*. Then, assuming $d_1 = d_2 = d_3 = d_4 = d$ and applying the Dirac interferometric principle

$$\langle d|s\rangle = \langle d|p_4\rangle\langle p_4|s\rangle + \langle d|p_3\rangle\langle p_3|s\rangle + \langle d|p_2\rangle\langle p_2|s\rangle + \langle d|p_1\rangle\langle p_1|s\rangle \qquad (18.2)$$

and following the procedure taught in chapter 17

$$|D\rangle_4 = |p_4\rangle\langle p_4|s\rangle \qquad (18.3)$$

$$|D\rangle_3 = |p_3\rangle\langle p_3|s\rangle \qquad (18.4)$$

$$|D\rangle_2 = |p_2\rangle\langle p_2|s\rangle \qquad (18.5)$$

$$|D\rangle_1 = |p_1\rangle\langle p_1|s\rangle \qquad (18.6)$$

so that equation (18.2) becomes

$$|s\rangle = (|D\rangle_4 + |D\rangle_3 + |D\rangle_2 + |D\rangle_1). \tag{18.7}$$

Again, applying the Dirac identity (Dirac, 1978), see chapter 17,

$$|X\rangle = |a\rangle_1 |b\rangle_2 |C\rangle_3...|g\rangle_n \tag{18.8}$$

and using $|C\rangle_m$'s to differentiate from a pure interferometric situation

$$|C\rangle_1 = |x\rangle_1 |y\rangle_2 |\varphi\rangle_3 |\varphi'\rangle_4 \tag{18.9}$$

$$|C\rangle_2 = |y\rangle_1 |x\rangle_2 |\varphi'\rangle_3 |\varphi\rangle_4 \tag{18.10}$$

$$|C\rangle_3 = |\varphi\rangle_1 |\varphi'\rangle_2 |x\rangle_3 |y\rangle_4 \tag{18.11}$$

$$|C\rangle_4 = |\varphi'\rangle_1 |\varphi\rangle_2 |y\rangle_3 |x\rangle_4. \tag{18.12}$$

Next, introducing the normalized probability amplitude $|\psi\rangle_R$, where R is a *Roman* numeral, the relation between $|s\rangle$ and $|\psi\rangle_R$ is given by (Duarte 2015)

$$|\psi\rangle_R = \frac{1}{\sqrt{N}}|s\rangle \tag{18.13}$$

and the normalization condition for $N = 4$ is given by

$$1 = \||\psi\rangle_I|^2 + \||\psi\rangle_{II}|^2 + \||\psi\rangle_{III}|^2 + \||\psi\rangle_{IV}|^2. \tag{18.14}$$

The normalization condition leads to

$$|\psi\rangle_I = \frac{1}{\sqrt{4}}(|C\rangle_1 + |C\rangle_2 + |C\rangle_3 + |C\rangle_4) \tag{18.15}$$

$$|\psi\rangle_{II} = \frac{1}{\sqrt{4}}(|C\rangle_1 + |C\rangle_2 - |C\rangle_3 - |C\rangle_4) \tag{18.16}$$

$$|\psi\rangle_{III} = \frac{1}{\sqrt{4}}(|C\rangle_1 - |C\rangle_2 + |C\rangle_3 - |C\rangle_4) \tag{18.17}$$

$$|\psi\rangle_{IV} = \frac{1}{\sqrt{4}}(|C\rangle_1 - |C\rangle_2 - |C\rangle_3 + |C\rangle_4). \tag{18.18}$$

Thus, putting identities (18.8)–(18.12) into equations (18.15)–(18.18), the explicit expressions for the probability amplitudes become

$$|\psi\rangle_I = \frac{1}{\sqrt{4}}(|x\rangle_1 |y\rangle_2 |\varphi\rangle_3 |\varphi'\rangle_4 + |y\rangle_1 |x\rangle_2 |\varphi'\rangle_3 |\varphi\rangle_4$$
$$+ |\varphi\rangle_1 |\varphi'\rangle_2 |x\rangle_3 |y\rangle_4 + |\varphi'\rangle_1 |\varphi\rangle_2 |y\rangle_3 |x\rangle_4) \tag{18.19}$$

$$|\psi\rangle_{II} = \frac{1}{\sqrt{4}}(|x\rangle_1 |y\rangle_2 |\varphi\rangle_3 |\varphi'\rangle_4 + |y\rangle_1 |x\rangle_2 |\varphi'\rangle_3 |\varphi\rangle_4$$
$$- |\varphi\rangle_1 |\varphi'\rangle_2 |x\rangle_3 |y\rangle_4 + |\varphi'\rangle_1 |\varphi\rangle_2 |y\rangle_3 |x\rangle_4) \tag{18.20}$$

$$|\psi\rangle_{III} = \frac{1}{\sqrt{4}}(|x\rangle_1 |y\rangle_2 |\varphi\rangle_3 |\varphi'\rangle_4 - |y\rangle_1 |x\rangle_2 |\varphi'\rangle_3 |\varphi\rangle_4$$
$$+ |\varphi\rangle_1 |\varphi'\rangle_2 |x\rangle_3 |y\rangle_4 - |\varphi'\rangle_1 |\varphi\rangle_2 |y\rangle_3 |x\rangle_4) \tag{18.21}$$

$$|\psi\rangle_{IV} = \frac{1}{\sqrt{4}}(|x\rangle_1 |y\rangle_2 |\varphi\rangle_3 |\varphi'\rangle_4 - |y\rangle_1 |x\rangle_2 |\varphi'\rangle_3 |\varphi\rangle_4$$
$$- |\varphi\rangle_1 |\varphi'\rangle_2 |x\rangle_3 |y\rangle_4 + |\varphi'\rangle_1 |\varphi\rangle_2 |y\rangle_3 |x\rangle_4). \tag{18.22}$$

It should be noted that the symmetry pattern in the sign convention of the $|C\rangle_m$ states can be best appreciated in a 4×4 arrangement (Duarte 2018):

$$+ |C\rangle_1 + |C\rangle_2 + |C\rangle_3 + |C\rangle_4$$
$$+ |C\rangle_1 + |C\rangle_2 - |C\rangle_3 - |C\rangle_4$$
$$+ |C\rangle_1 - |C\rangle_2 + |C\rangle_3 - |C\rangle_4 \tag{18.23}$$
$$+ |C\rangle_1 - |C\rangle_2 - |C\rangle_3 + |C\rangle_4.$$

18.3 The quantum entanglement probability amplitude for $n = N = 8$

The experimental diagram applicable to the $n = N = 8$ configuration of quantum entanglement is depicted in figure 18.2. Four axes are involved, each adjacent axis at $\theta = \pi/4$ rad relative to each other. Since each axis has a + and a − direction, $N = 8$ channels of propagation are available. Here, s is the photon source of *four pairs* of quanta emitted in opposite directions relative to each other. Again, in order to simplify the notation and axes labeling, in reference to figure 18.2, axes labels are obviated and only polarization alternatives are assigned in pairs; in this case, (x, y) along the $+z$ and $-z$ directions, and (φ, φ') along the corresponding orthogonal direction. The first axis at $\theta = \pi/4$, relative to the z axis, involves the (ϕ, ϕ') polarizations, and the axis orthogonal to it involves the (ϑ, ϑ') polarizations, as illustrated in figure 18.2.

Following the same methodology as in chapter 17 and the previous section,

$$|s\rangle = |D\rangle_8 + |D\rangle_7 + |D\rangle_6 + |D\rangle_5 + |D\rangle_4 + |D\rangle_3 + |D\rangle_2 + |D\rangle_1. \tag{18.24}$$

Again, applying the Dirac identity (Dirac 1978), see chapter 17,

$$|X\rangle = |a\rangle_1 |b\rangle_2 |C\rangle_3...|g\rangle_n \tag{18.25}$$

and using $|C\rangle_m$'s to differentiate from a pure interferometric situation

$$|C\rangle_1 = |x\rangle_1 |y\rangle_2 |\varphi\rangle_3 |\varphi'\rangle_4 |\phi\rangle_5 |\phi'\rangle_6 |\vartheta\rangle_7 |\vartheta'\rangle_8 \tag{18.26}$$

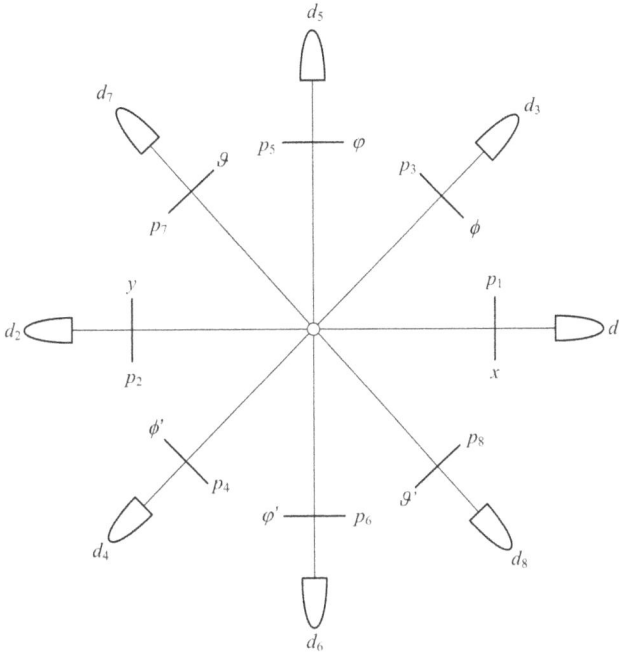

Figure 18.2. Experimental diagram applicable to quantum entanglement for $n = N = 2^3$.

$$|C\rangle_2 = |y\rangle_1 \, |x\rangle_2 \, |\varphi'\rangle_3 \, |\varphi\rangle_4 \, |\phi'\rangle_5 \, |\phi\rangle_6 \, |\vartheta'\rangle_7 \, |\vartheta\rangle_8 \tag{18.27}$$

$$|C\rangle_3 = |\varphi\rangle_1 \, |\varphi'\rangle_2 \, |\phi\rangle_3 \, |\phi'\rangle_4 \, |\vartheta\rangle_5 \, |\vartheta'\rangle_6 \, |x\rangle_7 \, |y\rangle_8 \tag{18.28}$$

$$|C\rangle_4 = |\varphi'\rangle_1 \, |\varphi\rangle_2 \, |\phi'\rangle_3 \, |\phi\rangle_4 \, |\vartheta'\rangle_5 \, |\vartheta\rangle_6 \, |y\rangle_7 \, |x\rangle_8 \tag{18.29}$$

$$|C\rangle_5 = |\phi\rangle_1 \, |\phi'\rangle_2 \, |\vartheta\rangle_3 \, |\vartheta'\rangle_4 \, |x\rangle_5 \, |y\rangle_6 \, |\varphi\rangle_7 \, |\varphi'\rangle_8 \tag{18.30}$$

$$|C\rangle_6 = |\phi'\rangle_1 \, |\phi\rangle_2 \, |\vartheta'\rangle_3 \, |\vartheta\rangle_4 \, |y\rangle_5 \, |x\rangle_6 \, |\varphi'\rangle_7 \, |\varphi\rangle_8 \tag{18.31}$$

$$|C\rangle_7 = |\vartheta\rangle_1 \, |\vartheta'\rangle_2 \, |x\rangle_3 \, |y\rangle_4 \, |\varphi\rangle_5 \, |\varphi'\rangle_6 \, |\phi\rangle_7 \, |\phi'\rangle_8 \tag{18.32}$$

$$|C\rangle_8 = |\vartheta'\rangle_1 \, |\vartheta\rangle_2 \, |y\rangle_3 \, |x\rangle_4 \, |\varphi'\rangle_5 \, |\varphi\rangle_6 \, |\phi'\rangle_7 \, |\phi\rangle_8. \tag{18.33}$$

Equations (18.26)–(18.33) describe quanta pairs (1, 2), (3, 4), (5, 6), (7, 8) in a particular polarization order as outlined by $|C\rangle_1$. Next, within pairs, polarizations are interchanged as given in $|C\rangle_2$. In $|C\rangle_3$, the second pair moves to the position of the first pair which is displaced to the position of the last pair. Transposition of the polarizations follows as indicated in $|C\rangle_4$, and so on (Duarte 2018). The only constant is the position of the quanta 1, 2, 3, 4, 5, 6, 7, 8.

For $n = N = 2^3$, the normalized probability amplitudes become

$$|\psi\rangle_I = \frac{1}{\sqrt{8}}(|C\rangle_1 + |C\rangle_2 + |C\rangle_3 + |C\rangle_4 + |C\rangle_5 + |C\rangle_6 + |C\rangle_7 + |C\rangle_8) \tag{18.34}$$

$$|\psi\rangle_{II} = \frac{1}{\sqrt{8}}(|C\rangle_1 + |C\rangle_2 + |C\rangle_3 + |C\rangle_4 - |C\rangle_5 - |C\rangle_6 - |C\rangle_7 - |C\rangle_8) \quad (18.35)$$

$$|\psi\rangle_{III} = \frac{1}{\sqrt{8}}(|C\rangle_1 + |C\rangle_2 - |C\rangle_3 - |C\rangle_4 + |C\rangle_5 + |C\rangle_6 - |C\rangle_7 - |C\rangle_8) \quad (18.36)$$

$$|\psi\rangle_{IV} = \frac{1}{\sqrt{8}}(|C\rangle_1 + |C\rangle_2 - |C\rangle_3 - |C\rangle_4 - |C\rangle_5 - |C\rangle_6 + |C\rangle_7 + |C\rangle_8) \quad (18.37)$$

$$|\psi\rangle_{V} = \frac{1}{\sqrt{8}}(|C\rangle_1 - |C\rangle_2 + |C\rangle_3 - |C\rangle_4 + |C\rangle_5 - |C\rangle_6 + |C\rangle_7 - |C\rangle_8) \quad (18.38)$$

$$|\psi\rangle_{VI} = \frac{1}{\sqrt{8}}(|C\rangle_1 - |C\rangle_2 + |C\rangle_3 - |C\rangle_4 - |C\rangle_5 + |C\rangle_6 - |C\rangle_7 + |C\rangle_8) \quad (18.39)$$

$$|\psi\rangle_{VII} = \frac{1}{\sqrt{8}}(|C\rangle_1 - |C\rangle_2 - |C\rangle_3 + |C\rangle_4 + |C\rangle_5 - |C\rangle_6 - |C\rangle_7 + |C\rangle_8) \quad (18.40)$$

$$|\psi\rangle_{VIII} = \frac{1}{\sqrt{8}}(|C\rangle_1 - |C\rangle_2 - |C\rangle_3 + |C\rangle_4 - |C\rangle_5 + |C\rangle_6 + |C\rangle_7 - |C\rangle_8). \quad (18.41)$$

For the row and column originating at $+C_1$, the sign sequences (right and down) are +, +, +, +, +, +, +, + and +, +, +, +, +, +, +, +. On the other hand, for the terms on the diagonal the sign sequence is +, +, −, −, +, +, −, −. The sign sequence for the second column and the second row is +, +, +, +, −, −, −, − while the sequence for the third column and the third row is +, +, −, −, +, +, −, −, and so on.

A better appreciation of the sign alternatives is offered if the $|C\rangle_m$ amplitudes are expressed in an 8×8 format with the *bra–ket* ($|\ \rangle$) symbols abstracted, so that (Duarte 2018)

$$
\begin{aligned}
&+ C_1 + C_2 + C_3 + C_4 + C_5 + C_6 + C_7 + C_8 \\
&+ C_1 + C_2 + C_3 + C_4 - C_5 - C_6 - C_7 - C_8 \\
&+ C_1 + C_2 - C_3 - C_4 + C_5 + C_6 - C_7 - C_8 \\
&+ C_1 + C_2 - C_3 - C_4 - C_5 - C_6 + C_7 + C_8 \\
&+ C_1 - C_2 + C_3 - C_4 + C_5 - C_6 + C_7 - C_8 \\
&+ C_1 - C_2 + C_3 - C_4 - C_5 + C_6 - C_7 + C_8 \\
&+ C_1 - C_2 - C_3 + C_4 + C_5 - C_6 - C_7 + C_8 \\
&+ C_1 - C_2 - C_3 + C_4 - C_5 + C_6 + C_7 - C_8.
\end{aligned}
\quad (18.42)
$$

18.4 The quantum entanglement probability amplitude for $n = N = 16$

For $n = N = 2^4$, the first normalized probability $|\psi\rangle_I$ amplitudes become

$$
\begin{aligned}
|\psi\rangle_I = \frac{1}{\sqrt{16}}(&|C\rangle_1 + |C\rangle_2 + |C\rangle_3 + |C\rangle_4 + |C\rangle_5 + |C\rangle_6 + |C\rangle_7 + |C\rangle_8 \\
&+ |C\rangle_9 + |C\rangle_{10} + |C\rangle_{11} + |C\rangle_{12} + |C\rangle_{13} + |C\rangle_{14} + |C\rangle_{15} + |C\rangle_{16})
\end{aligned}
\tag{18.43}
$$

and so on for $|\psi\rangle_{II}, |\psi\rangle_{III}, |\psi\rangle_{IV} \dots |\psi\rangle_{XVI}$. Going straight to the sign alternatives for the $|C\rangle_m$ probability amplitudes expressed in an 16×16 format, with the *bra–ket* $(|\;\rangle)$ symbols abstracted (Duarte 2018),

$$
\begin{aligned}
&+ C_1 + C_2 + C_3 + C_4 + C_5 + C_6 + C_7 + C_8 + C_9 + C_{10} + C_{11} + C_{12} + C_{13} + C_{14} + C_{15} + C_{16} \\
&+ C_1 + C_2 + C_3 + C_4 + C_5 + C_6 + C_7 + C_8 - C_9 - C_{10} - C_{11} - C_{12} - C_{13} - C_{14} - C_{15} - C_{16} \\
&+ C_1 + C_2 + C_3 + C_4 - C_5 - C_6 - C_7 - C_8 + C_9 + C_{10} + C_{11} + C_{12} - C_{13} - C_{14} - C_{15} - C_{16} \\
&+ C_1 + C_2 + C_3 + C_4 - C_5 - C_6 - C_7 - C_8 - C_9 - C_{10} - C_{11} - C_{12} + C_{13} + C_{14} + C_{15} + C_{16} \\
&+ C_1 + C_2 - C_3 - C_4 + C_5 + C_6 - C_7 - C_8 + C_9 + C_{10} - C_{11} - C_{12} + C_{13} + C_{14} - C_{15} - C_{16} \\
&+ C_1 + C_2 - C_3 - C_4 + C_5 + C_6 - C_7 - C_8 - C_9 - C_{10} + C_{11} + C_{12} - C_{13} - C_{14} + C_{15} + C_{16} \\
&+ C_1 + C_2 - C_3 - C_4 - C_5 - C_6 + C_7 + C_8 + C_9 + C_{10} - C_{11} - C_{12} - C_{13} - C_{14} + C_{15} + C_{16} \\
&+ C_1 + C_2 - C_3 - C_4 - C_5 - C_6 + C_7 + C_8 - C_9 - C_{10} + C_{11} + C_{12} + C_{13} + C_{14} - C_{15} - C_{16} \\
&+ C_1 - C_2 + C_3 - C_4 + C_5 - C_6 + C_7 - C_8 + C_9 - C_{10} + C_{11} - C_{12} + C_{13} - C_{14} + C_{15} - C_{16} \\
&+ C_1 - C_2 + C_3 - C_4 + C_5 - C_6 + C_7 - C_8 - C_9 + C_{10} - C_{11} + C_{12} - C_{13} + C_{14} - C_{15} + C_{16} \\
&+ C_1 - C_2 + C_3 - C_4 - C_5 + C_6 - C_7 + C_8 + C_9 - C_{10} + C_{11} - C_{12} - C_{13} + C_{14} - C_{15} + C_{16} \\
&+ C_1 - C_2 + C_3 - C_4 - C_5 + C_6 - C_7 + C_8 - C_9 + C_{10} - C_{11} + C_{12} + C_{13} - C_{14} + C_{15} - C_{16} \\
&+ C_1 - C_2 - C_3 + C_4 + C_5 - C_6 - C_7 + C_8 + C_9 - C_{10} - C_{11} + C_{12} + C_{13} - C_{14} - C_{15} + C_{16} \\
&+ C_1 - C_2 - C_3 + C_4 + C_5 - C_6 - C_7 + C_8 - C_9 + C_{10} + C_{11} - C_{12} - C_{13} + C_{14} + C_{15} - C_{16} \\
&+ C_1 - C_2 - C_3 + C_4 - C_5 + C_6 + C_7 - C_8 + C_9 - C_{10} - C_{11} + C_{12} - C_{13} + C_{14} + C_{15} - C_{16} \\
&+ C_1 - C_2 - C_3 + C_4 - C_5 + C_6 + C_7 - C_8 - C_9 + C_{10} + C_{11} - C_{12} + C_{13} - C_{14} - C_{15} + C_{16}.
\end{aligned}
\tag{18.44}
$$

18.5 The quantum entanglement probability amplitude for $n = N = 2^1, 2^2, 2^3, 2^4 \dots 2^r$

First, the probability amplitude $|s\rangle$ can generally be expressed as

$$
|s\rangle = \sum_{j=1}^{N} (\pm)|C\rangle_{N+1-j}
\tag{18.45}
$$

and for notational consistency the normalized probability amplitude is expressed as

$$
|\psi\rangle_R = \frac{1}{\sqrt{N}}|s\rangle.
\tag{18.46}
$$

This means that

$$
|\psi\rangle_R = \frac{1}{\sqrt{N}}\sum_{j=1}^{N} (\pm)|C\rangle_{N+1-j}
\tag{18.47}
$$

where the subscript R is a Roman numeral, the $1/\sqrt{N}$ factor is derived from the normalization operation

$$1 = \||\psi\rangle_I|^2 + \||\psi\rangle_{II}|^2 + \||\psi\rangle_{III}|^2 + \||\psi\rangle_{IV}|^2 + \cdots \qquad (18.48)$$

and the sign alternative ($\pm$) means that every possible sign alternative in the series is allowed.

Next, using the Dirac's identity for 'similar particles', the individual path probability amplitudes $|C\rangle_{N+1-j}$ are expressed in a generalized form as (Duarte 2016, 2018)

$$|C\rangle_{N+1-j} = \prod_{m=1,3,5...}^{n} |a\rangle_m |b\rangle_{m+1}. \qquad (18.49)$$

Equation (18.49) is a mathematical expression for a series of probability amplitudes beginning at $|C\rangle_{N+1-j}$ with $j = 1$ and ending at $|C\rangle_1$, which is reached when $j = N$. In this notation, n is the total number of quanta and is an even number since quanta participate in pairs. For each pair, $(1, 2), (3, 4), \ldots, (m, m + 1)$, and $|a\rangle_m, |a\rangle_{m+1}$ represent a set of orthogonal polarization alternatives such as $(x, y), (\varphi, \varphi')$, and so on (Duarte 2018).

Explicitly, for $n = N = 2^3$, and for the values of j expressed in parentheses, equation (18.49) yields

$(j = N - 0)$:
$$|C\rangle_1 = |x\rangle_1 |y\rangle_2 |\varphi\rangle_3 |\varphi'\rangle_4 |\phi\rangle_5 |\phi'\rangle_6 |\vartheta\rangle_7 |\vartheta'\rangle_8 \qquad (18.50)$$

$(j = N - 1)$:
$$|C\rangle_2 = |y\rangle_1 |x\rangle_2 |\varphi'\rangle_3 |\varphi\rangle_4 |\phi'\rangle_5 |\phi\rangle_6 |\vartheta'\rangle_7 |\vartheta\rangle_8 \qquad (18.51)$$

$(j = N - 2)$:
$$|C\rangle_3 = |\varphi\rangle_1 |\varphi'\rangle_2 |\phi\rangle_3 |\phi'\rangle_4 |\vartheta\rangle_5 |\vartheta'\rangle_6 |x\rangle_7 |y\rangle_8 \qquad (18.52)$$

$(j = N - 3)$:
$$|C\rangle_4 = |\varphi'\rangle_1 |\varphi\rangle_2 |\phi'\rangle_3 |\phi\rangle_4 |\vartheta'\rangle_5 |\vartheta\rangle_6 |y\rangle_7 |x\rangle_8 \qquad (18.53)$$

$(j = N - 4)$:
$$|C\rangle_5 = |\phi\rangle_1 |\phi'\rangle_2 |\vartheta\rangle_3 |\vartheta'\rangle_4 |x\rangle_5 |y\rangle_6 |\varphi\rangle_7 |\varphi'\rangle_8 \qquad (18.54)$$

$(j = N - 5)$:
$$|C\rangle_6 = |\phi'\rangle_1 |\phi\rangle_2 |\vartheta'\rangle_3 |\vartheta\rangle_4 |y\rangle_5 |x\rangle_6 |\varphi'\rangle_7 |\varphi\rangle_8 \qquad (18.55)$$

$(j = N - 6)$:
$$|C\rangle_7 = |\vartheta\rangle_1 |\vartheta'\rangle_2 |x\rangle_3 |y\rangle_4 |\varphi\rangle_5 |\varphi'\rangle_6 |\phi\rangle_7 |\phi'\rangle_8 \qquad (18.56)$$

$(j = N - 7):$

$$|C\rangle_8 = |\vartheta'\rangle_1 |\vartheta\rangle_2 |y\rangle_3 |x\rangle_4 |\varphi'\rangle_5 |\varphi\rangle_6 |\phi'\rangle_7 |\phi\rangle_8.$$

(18.57)

Substitution of the corresponding $|C\rangle_{N+1-j}$ amplitudes, that is, $|C\rangle_1$, $|C\rangle_2$, $|C\rangle_3$, ..., $|C\rangle_8$ (from equations (18.50)–(18.57)), into equations (18.34)–(18.41), which are also directly obtainable from equation (18.47), leads to the explicit series of probability amplitudes $|\psi\rangle_I$, $|\psi\rangle_{II}$, $|\psi\rangle_{III}$, ..., $|\psi\rangle_{VIII}$ in terms of the polarization coordinates (x, y), (φ, φ'), (ϕ, ϕ'), and (ϑ, ϑ') (Duarte and Taylor 2017):

$$
\begin{aligned}
|\psi\rangle_I = \frac{1}{\sqrt{8}}(&|x\rangle_1 |y\rangle_2 |\varphi\rangle_3 |\varphi'\rangle_4 |\phi\rangle_5 |\phi'\rangle_6 |\vartheta\rangle_7 |\vartheta'\rangle_8 \\
+ &|y\rangle_1 |x\rangle_2 |\varphi'\rangle_3 |\varphi\rangle_4 |\phi'\rangle_5 |\phi\rangle_6 |\vartheta'\rangle_7 |\vartheta\rangle_8 \\
+ &|\varphi\rangle_1 |\varphi'\rangle_2 |\phi\rangle_3 |\phi'\rangle_4 |\vartheta\rangle_5 |\vartheta'\rangle_6 |x\rangle_7 |y\rangle_8 \\
+ &|\varphi'\rangle_1 |\varphi\rangle_2 |\phi'\rangle_3 |\phi\rangle_4 |\vartheta'\rangle_5 |\vartheta\rangle_6 |y\rangle_7 |x\rangle_8 \\
+ &|\phi\rangle_1 |\phi'\rangle_2 |\vartheta\rangle_3 |\vartheta'\rangle_4 |x\rangle_5 |y\rangle_6 |\varphi\rangle_7 |\varphi'\rangle_8 \\
+ &|\phi'\rangle_1 |\phi\rangle_2 |\vartheta'\rangle_3 |\vartheta\rangle_4 |y\rangle_5 |x\rangle_6 |\varphi'\rangle_7 |\varphi\rangle_8 \\
+ &|\vartheta\rangle_1 |\vartheta'\rangle_2 |x\rangle_3 |y\rangle_4 |\varphi\rangle_5 |\varphi'\rangle_6 |\phi\rangle_7 |\phi'\rangle_8 \\
+ &|\vartheta'\rangle_1 |\vartheta\rangle_2 |y\rangle_3 |x\rangle_4 |\varphi'\rangle_5 |\varphi\rangle_6 |\phi'\rangle_7 |\phi\rangle_8)
\end{aligned}
$$

(18.58)

$$
\begin{aligned}
|\psi\rangle_{II} = \frac{1}{\sqrt{8}}(&|x\rangle_1 |y\rangle_2 |\varphi\rangle_3 |\varphi'\rangle_4 |\phi\rangle_5 |\phi'\rangle_6 |\vartheta\rangle_7 |\vartheta'\rangle_8 \\
+ &|y\rangle_1 |x\rangle_2 |\varphi'\rangle_3 |\varphi\rangle_4 |\phi'\rangle_5 |\phi\rangle_6 |\vartheta'\rangle_7 |\vartheta\rangle_8 \\
+ &|\varphi\rangle_1 |\varphi'\rangle_2 |\phi\rangle_3 |\phi'\rangle_4 |\vartheta\rangle_5 |\vartheta'\rangle_6 |x\rangle_7 |y\rangle_8 \\
+ &|\varphi'\rangle_1 |\varphi\rangle_2 |\phi'\rangle_3 |\phi\rangle_4 |\vartheta'\rangle_5 |\vartheta\rangle_6 |y\rangle_7 |x\rangle_8 \\
- &|\phi\rangle_1 |\phi'\rangle_2 |\vartheta\rangle_3 |\vartheta'\rangle_4 |x\rangle_5 |y\rangle_6 |\varphi\rangle_7 |\varphi'\rangle_8 \\
- &|\phi'\rangle_1 |\phi\rangle_2 |\vartheta'\rangle_3 |\vartheta\rangle_4 |y\rangle_5 |x\rangle_6 |\varphi'\rangle_7 |\varphi\rangle_8 \\
- &|\vartheta\rangle_1 |\vartheta'\rangle_2 |x\rangle_3 |y\rangle_4 |\varphi\rangle_5 |\varphi'\rangle_6 |\phi\rangle_7 |\phi'\rangle_8 \\
- &|\vartheta'\rangle_1 |\vartheta\rangle_2 |y\rangle_3 |x\rangle_4 |\varphi'\rangle_5 |\varphi\rangle_6 |\phi'\rangle_7 |\phi\rangle_8)
\end{aligned}
$$

(18.59)

$$
\begin{aligned}
|\psi\rangle_{III} = \frac{1}{\sqrt{8}}(&|x\rangle_1 |y\rangle_2 |\varphi\rangle_3 |\varphi'\rangle_4 |\phi\rangle_5 |\phi'\rangle_6 |\vartheta\rangle_7 |\vartheta'\rangle_8 \\
+ &|y\rangle_1 |x\rangle_2 |\varphi'\rangle_3 |\varphi\rangle_4 |\phi'\rangle_5 |\phi\rangle_6 |\vartheta'\rangle_7 |\vartheta\rangle_8 \\
- &|\varphi\rangle_1 |\varphi'\rangle_2 |\phi\rangle_3 |\phi'\rangle_4 |\vartheta\rangle_5 |\vartheta'\rangle_6 |x\rangle_7 |y\rangle_8 \\
- &|\varphi'\rangle_1 |\varphi\rangle_2 |\phi'\rangle_3 |\phi\rangle_4 |\vartheta'\rangle_5 |\vartheta\rangle_6 |y\rangle_7 |x\rangle_8 \\
+ &|\phi\rangle_1 |\phi'\rangle_2 |\vartheta\rangle_3 |\vartheta'\rangle_4 |x\rangle_5 |y\rangle_6 |\varphi\rangle_7 |\varphi'\rangle_8 \\
+ &|\phi'\rangle_1 |\phi\rangle_2 |\vartheta'\rangle_3 |\vartheta\rangle_4 |y\rangle_5 |x\rangle_6 |\varphi'\rangle_7 |\varphi\rangle_8 \\
- &|\vartheta\rangle_1 |\vartheta'\rangle_2 |x\rangle_3 |y\rangle_4 |\varphi\rangle_5 |\varphi'\rangle_6 |\phi\rangle_7 |\phi'\rangle_8 \\
- &|\vartheta'\rangle_1 |\vartheta\rangle_2 |y\rangle_3 |x\rangle_4 |\varphi'\rangle_5 |\varphi\rangle_6 |\phi'\rangle_7 |\phi\rangle_8)
\end{aligned}
$$

(18.60)

$$|\psi\rangle_{IV} = \frac{1}{\sqrt{8}}(|x\rangle_1 |y\rangle_2 |\varphi\rangle_3 |\varphi'\rangle_4 |\phi\rangle_5 |\phi'\rangle_6 |\vartheta\rangle_7 |\vartheta'\rangle_8$$
$$+ |y\rangle_1 |x\rangle_2 |\varphi'\rangle_3 |\varphi\rangle_4 |\phi'\rangle_5 |\phi\rangle_6 |\vartheta'\rangle_7 |\vartheta\rangle_8$$
$$- |\varphi\rangle_1 |\varphi'\rangle_2 |\phi\rangle_3 |\phi'\rangle_4 |\vartheta\rangle_5 |\vartheta'\rangle_6 |x\rangle_7 |y\rangle_8$$
$$- |\varphi'\rangle_1 |\varphi\rangle_2 |\phi'\rangle_3 |\phi\rangle_4 |\vartheta'\rangle_5 |\vartheta\rangle_6 |y\rangle_7 |x\rangle_8 \qquad (18.61)$$
$$- |\phi\rangle_1 |\phi'\rangle_2 |\vartheta\rangle_3 |\vartheta'\rangle_4 |x\rangle_5 |y\rangle_6 |\varphi\rangle_7 |\varphi'\rangle_8$$
$$- |\phi'\rangle_1 |\phi\rangle_2 |\vartheta'\rangle_3 |\vartheta\rangle_4 |y\rangle_5 |x\rangle_6 |\varphi'\rangle_7 |\varphi\rangle_8$$
$$+ |\vartheta\rangle_1 |\vartheta'\rangle_2 |x\rangle_3 |y\rangle_4 |\varphi\rangle_5 |\varphi'\rangle_6 |\phi\rangle_7 |\phi'\rangle_8$$
$$+ |\vartheta'\rangle_1 |\vartheta\rangle_2 |y\rangle_3 |x\rangle_4 |\varphi'\rangle_5 |\varphi\rangle_6 |\phi'\rangle_7 |\phi\rangle_8)$$

$$|\psi\rangle_V = \frac{1}{\sqrt{8}}(|x\rangle_1 |y\rangle_2 |\varphi\rangle_3 |\varphi'\rangle_4 |\phi\rangle_5 |\phi'\rangle_6 |\vartheta\rangle_7 |\vartheta'\rangle_8$$
$$- |y\rangle_1 |x\rangle_2 |\varphi'\rangle_3 |\varphi\rangle_4 |\phi'\rangle_5 |\phi\rangle_6 |\vartheta'\rangle_7 |\vartheta\rangle_8$$
$$+ |\varphi\rangle_1 |\varphi'\rangle_2 |\phi\rangle_3 |\phi'\rangle_4 |\vartheta\rangle_5 |\vartheta'\rangle_6 |x\rangle_7 |y\rangle_8$$
$$- |\varphi'\rangle_1 |\varphi\rangle_2 |\phi'\rangle_3 |\phi\rangle_4 |\vartheta'\rangle_5 |\vartheta\rangle_6 |y\rangle_7 |x\rangle_8 \qquad (18.62)$$
$$+ |\phi\rangle_1 |\phi'\rangle_2 |\vartheta\rangle_3 |\vartheta'\rangle_4 |x\rangle_5 |y\rangle_6 |\varphi\rangle_7 |\varphi'\rangle_8$$
$$- |\phi'\rangle_1 |\phi\rangle_2 |\vartheta'\rangle_3 |\vartheta\rangle_4 |y\rangle_5 |x\rangle_6 |\varphi'\rangle_7 |\varphi\rangle_8$$
$$+ |\vartheta\rangle_1 |\vartheta'\rangle_2 |x\rangle_3 |y\rangle_4 |\varphi\rangle_5 |\varphi'\rangle_6 |\phi\rangle_7 |\phi'\rangle_8$$
$$- |\vartheta'\rangle_1 |\vartheta\rangle_2 |y\rangle_3 |x\rangle_4 |\varphi'\rangle_5 |\varphi\rangle_6 |\phi'\rangle_7 |\phi\rangle_8)$$

$$|\psi\rangle_{VI} = \frac{1}{\sqrt{8}}(|x\rangle_1 |y\rangle_2 |\varphi\rangle_3 |\varphi'\rangle_4 |\phi\rangle_5 |\phi'\rangle_6 |\vartheta\rangle_7 |\vartheta'\rangle_8$$
$$- |y\rangle_1 |x\rangle_2 |\varphi'\rangle_3 |\varphi\rangle_4 |\phi'\rangle_5 |\phi\rangle_6 |\vartheta'\rangle_7 |\vartheta\rangle_8$$
$$+ |\varphi\rangle_1 |\varphi'\rangle_2 |\phi\rangle_3 |\phi'\rangle_4 |\vartheta\rangle_5 |\vartheta'\rangle_6 |x\rangle_7 |y\rangle_8$$
$$- |\varphi'\rangle_1 |\varphi\rangle_2 |\phi'\rangle_3 |\phi\rangle_4 |\vartheta'\rangle_5 |\vartheta\rangle_6 |y\rangle_7 |x\rangle_8 \qquad (18.63)$$
$$- |\phi\rangle_1 |\phi'\rangle_2 |\vartheta\rangle_3 |\vartheta'\rangle_4 |x\rangle_5 |y\rangle_6 |\varphi\rangle_7 |\varphi'\rangle_8$$
$$+ |\phi'\rangle_1 |\phi\rangle_2 |\vartheta'\rangle_3 |\vartheta\rangle_4 |y\rangle_5 |x\rangle_6 |\varphi'\rangle_7 |\varphi\rangle_8$$
$$- |\vartheta\rangle_1 |\vartheta'\rangle_2 |x\rangle_3 |y\rangle_4 |\varphi\rangle_5 |\varphi'\rangle_6 |\phi\rangle_7 |\phi'\rangle_8$$
$$+ |\vartheta'\rangle_1 |\vartheta\rangle_2 |y\rangle_3 |x\rangle_4 |\varphi'\rangle_5 |\varphi\rangle_6 |\phi'\rangle_7 |\phi\rangle_8)$$

$$|\psi\rangle_{VII} = \frac{1}{\sqrt{8}}(|x\rangle_1 |y\rangle_2 |\varphi\rangle_3 |\varphi'\rangle_4 |\phi\rangle_5 |\phi'\rangle_6 |\vartheta\rangle_7 |\vartheta'\rangle_8$$
$$- |y\rangle_1 |x\rangle_2 |\varphi'\rangle_3 |\varphi\rangle_4 |\phi'\rangle_5 |\phi\rangle_6 |\vartheta'\rangle_7 |\vartheta\rangle_8$$
$$- |\varphi\rangle_1 |\varphi'\rangle_2 |\phi\rangle_3 |\phi'\rangle_4 |\vartheta\rangle_5 |\vartheta'\rangle_6 |x\rangle_7 |y\rangle_8$$
$$+ |\varphi'\rangle_1 |\varphi\rangle_2 |\phi'\rangle_3 |\phi\rangle_4 |\vartheta'\rangle_5 |\vartheta\rangle_6 |y\rangle_7 |x\rangle_8 \qquad (18.64)$$
$$+ |\phi\rangle_1 |\phi'\rangle_2 |\vartheta\rangle_3 |\vartheta'\rangle_4 |x\rangle_5 |y\rangle_6 |\varphi\rangle_7 |\varphi'\rangle_8$$
$$- |\phi'\rangle_1 |\phi\rangle_2 |\vartheta'\rangle_3 |\vartheta\rangle_4 |y\rangle_5 |x\rangle_6 |\varphi'\rangle_7 |\varphi\rangle_8$$
$$- |\vartheta\rangle_1 |\vartheta'\rangle_2 |x\rangle_3 |y\rangle_4 |\varphi\rangle_5 |\varphi'\rangle_6 |\phi\rangle_7 |\phi'\rangle_8$$
$$+ |\vartheta'\rangle_1 |\vartheta\rangle_2 |y\rangle_3 |x\rangle_4 |\varphi'\rangle_5 |\varphi\rangle_6 |\phi'\rangle_7 |\phi\rangle_8)$$

$$|\psi\rangle_{VIII} = \frac{1}{\sqrt{8}}(|x\rangle_1 \, |y\rangle_2 \, |\varphi\rangle_3 \, |\varphi'\rangle_4 \, |\phi\rangle_5 \, |\phi'\rangle_6 \, |\vartheta\rangle_7 \, |\vartheta'\rangle_8$$

$$- |y\rangle_1 \, |x\rangle_2 \, |\varphi'\rangle_3 \, |\varphi\rangle_4 \, |\phi'\rangle_5 \, |\phi\rangle_6 \, |\vartheta'\rangle_7 \, |\vartheta\rangle_8$$

$$- |\varphi\rangle_1 \, |\varphi'\rangle_2 \, |\phi\rangle_3 \, |\phi'\rangle_4 \, |\vartheta\rangle_5 \, |\vartheta'\rangle_6 \, |x\rangle_7 \, |y\rangle_8$$

$$+ |\varphi'\rangle_1 \, |\varphi\rangle_2 \, |\phi'\rangle_3 \, |\phi\rangle_4 \, |\vartheta'\rangle_5 \, |\vartheta\rangle_6 \, |y\rangle_7 \, |x\rangle_8 \tag{18.65}$$

$$- |\phi\rangle_1 \, |\phi'\rangle_2 \, |\vartheta\rangle_3 \, |\vartheta'\rangle_4 \, |x\rangle_5 \, |y\rangle_6 \, |\varphi\rangle_7 \, |\varphi'\rangle_8$$

$$+ |\phi'\rangle_1 \, |\phi\rangle_2 \, |\vartheta'\rangle_3 \, |\vartheta\rangle_4 \, |y\rangle_5 \, |x\rangle_6 \, |\varphi'\rangle_7 \, |\varphi\rangle_8$$

$$+ |\vartheta\rangle_1 \, |\vartheta'\rangle_2 \, |x\rangle_3 \, |y\rangle_4 \, |\varphi\rangle_5 \, |\varphi'\rangle_6 \, |\phi\rangle_7 \, |\phi'\rangle_8$$

$$- |\vartheta'\rangle_1 \, |\vartheta\rangle_2 \, |y\rangle_3 \, |x\rangle_4 \, |\varphi'\rangle_5 \, |\varphi\rangle_6 \, |\phi'\rangle_7 \, |\phi\rangle_8).$$

18.6 Discussion

Examination of quantum entanglement involving n quanta and N propagation channels in $n = N = 2^1, 2^2, 2^3 \ldots 2^r$ configurations leads to the following conclusions:

1. Each individual probability amplitude includes $n = N$ terms. See, for example, equations (18.19)–(18.22) or equations (18.34)–(18.41).
2. The normalization factor for a quantum entanglement configuration including N propagation channels is $1/\sqrt{N}$.
3. For a given $n = N$ entanglement configuration there are $n = N$ probability amplitudes obeying sign permutations leading to symmetric sign distributions starting with all + signs for the first row and the first column of the arrangement. For instance, for $n = N = 2^2$ and $n = N = 2^4$ the sign distributions for the first row, the first column, and the diagonal are +, +, +, + and +, +, +, +, +, +, +, +, +, +, +, +, +, +, +, respectively. However, for $n = N = 2^3$ the sign distribution for the first row and the first column are +, +, +, +, +, +, + while that for the diagonal is +, +, −, −, +, +, −, −.

By utilizing

$$|\psi\rangle_R = \frac{1}{\sqrt{N}} \sum_{j=1}^{N} (\pm)|C\rangle_{N+1-j}$$

$$1 = \||\psi\rangle_I\|^2 + \||\psi\rangle_{II}\|^2 + \||\psi\rangle_{III}\|^2 + \||\psi\rangle_{IV}\|^2 + \cdots$$

and

$$|C\rangle_{N+1-j} = \prod_{m=1,3,5\ldots}^{n} |a\rangle_m \, |b\rangle_{m+1}$$

the calculation of probability amplitudes for $n = N = 2^1, 2^2, 2^3, 2^4 \ldots 2^r$ configurations becomes a mechanical process.

References

Dirac P A M 1978 *The Principles of Quantum Mechanics* 4th edn. (Oxford: Oxford University Press)

Duarte F J 2013a The probability amplitude for entangled polarizations: an interferometric approach *J. Mod. Opt.* **60** 1585–7

Duarte F J 2013b Tunable laser optics: applications to optics and quantum optics *Prog. Quantum Electron.* **37** 326–47

Duarte F J 2014 *Quantum Optics for Engineers* (New York: CRC)

Duarte F J 2015 *Tunable Laser Optics* 2nd edn. (New York: CRC)

Duarte F J 2016 Secure space-to-space interferometric communications and its nexus to the physics of quantum entanglement *Appl. Phys. Rev.* **3** 041301

Duarte F J 2018 Organic lasers for *N*-channel quantum entanglement *Organic Lasers and Organic Photonics* ed F J Duarte (Bristol: IOP Publishing) ch 15

Duarte F J and Taylor T S 2017 Quantum entanglement probability amplitudes in multiple channels: an interferometric approach *Optik* **139** 222–30

IOP Publishing

Fundamentals of Quantum Entanglement

F J Duarte

Chapter 19

The interferometric derivation of the quantum entanglement probability amplitudes for $n = N = 3, 6$

The interferometric derivation of probability amplitudes for $n = N = 3, 6$ is described in detail using the Dirac–Feynman principle and Dirac's identities. This approach requires the use of Hamilton's quaternions. Here, n is the number of quanta and N is the number of propagation channels.

19.1 Introduction

In chapter 17 the interferometric derivation, *à la Dirac*, of the probability amplitude for quantum entanglement applicable to two quanta ($n = 2$) and two propagation channels ($N = 2$) was described. In chapter 18 the same derivational approach utilized in chapter 17 was applied to the cases of $n = N = 4, n = N = 8, n = N = 16$, and in general $n = N = 2^1, 2^2, 2^3, 2^4, ..., 2^r$, where $r = 1, 2, 3, 4, 5, ...$

In this chapter the interferometric derivation, *à la Dirac*, of the probability amplitude for quantum entanglement applicable to three quanta ($n = 3$) and three propagation channels ($N = 3$) is described. The same approach is then extended to $n = N = 6$.

The material presented in this chapter is a summary and extension of previous publications (Duarte 2015, 2016).

19.2 The quantum entanglement probability amplitude for $n = N = 3$

The experimental diagram applicable to the $n = N = 3$ configuration for quantum entanglement is depicted in figure 19.1. Here, s is the photon source of three quanta emitted in three directions at a relative angle of $2\pi/3$ (although this orientation is not mandatory).

doi:10.1088/2053-2563/ab2b33ch19

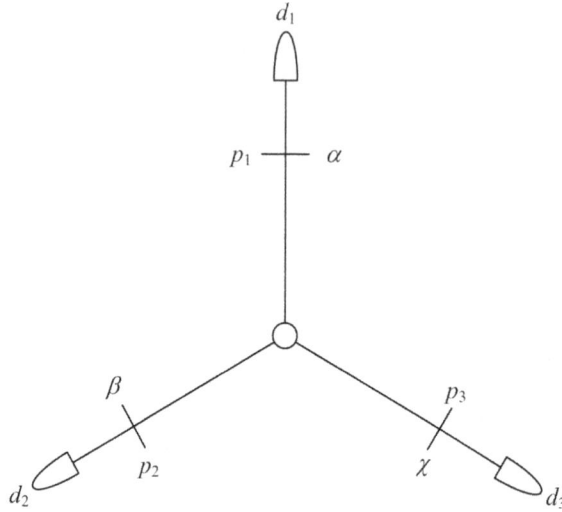

Figure 19.1. Experimental diagram applicable to quantum entanglement for the $n = N = 3$ configuration.

Again, the first essential assumption is that the quanta emitted by the source are *indistinguishable*. Then, assuming $d_1 = d_2 = d_3 = d$ and applying the Dirac interferometric principle, the probability amplitude

$$\langle d|s \rangle = \langle d|p_3 \rangle \langle p_3|s \rangle + \langle d|p_2 \rangle \langle p_2|s \rangle + \langle d|p_1 \rangle \langle p_1|s \rangle \qquad (19.1)$$

is obtained, and following the procedure taught in chapter 17

$$|D\rangle_3 = |p_3\rangle\langle p_3|s\rangle \qquad (19.2)$$

$$|D\rangle_2 = |p_2\rangle\langle p_2|s\rangle \qquad (19.3)$$

$$|D\rangle_1 = |p_1\rangle\langle p_1|s\rangle \qquad (19.4)$$

so that equation (19.1) becomes

$$|s\rangle = (|D\rangle_3 + |D\rangle_2 + |D\rangle_1). \qquad (19.5)$$

Again, applying the Dirac identity (Dirac 1978), see chapter 17,

$$|X\rangle = |a\rangle_1|b\rangle_2|c\rangle_3... |g\rangle_n \qquad (19.6)$$

and using $|C\rangle_m$'s to differentiate from a pure interferometric situation (see figure 19.1),

$$|C\rangle_1 = |a\rangle_1|\beta\rangle_2|\chi\rangle_3 \qquad (19.7)$$

$$|C\rangle_2 = |\chi\rangle_1|a\rangle_2|\beta\rangle_3 \qquad (19.8)$$

$$|C\rangle_3 = |\beta\rangle_1|\chi\rangle_2|a\rangle_3. \qquad (19.9)$$

At this stage the generalized probability amplitude introduced in chapter 18

$$|\psi\rangle_R = \frac{1}{\sqrt{N}} \sum_{j=1}^{N} (\pm)|C\rangle_{N+1-j} \qquad (19.10)$$

would be applied. This is carried out as previously but with one caveat: there is a need to introduce Hamilton's quaternions (Hamilton 1866) (see appendix J) in order to satisfy the normalization condition

$$1 = ||\psi\rangle_I|^2 + ||\psi\rangle_{II}|^2 + ||\psi\rangle_{III}|^2. \qquad (19.11)$$

This means that the probability amplitudes become

$$|\psi\rangle_I = \frac{1}{\sqrt{3}}(|C\rangle_1 + |C\rangle_2 + i\,|C\rangle_3) \qquad (19.12)$$

$$|\psi\rangle_{II} = \frac{1}{\sqrt{3}}(|C\rangle_1 - |C\rangle_2 + i\,|C\rangle_3) \qquad (19.13)$$

$$|\psi\rangle_{III} = \frac{1}{\sqrt{3}}(i\,|C\rangle_1 + j\,|C\rangle_2 + k\,|C\rangle_3). \qquad (19.14)$$

Thus, by substituting equations (19.7)–(19.9) into equations (19.12)–(19.14), the explicit probability amplitudes are revealed:

$$|\psi\rangle_I = \frac{1}{\sqrt{3}}(|\alpha\rangle_1|\beta\rangle_2|\chi\rangle_3 + |\chi\rangle_1|\alpha\rangle_2|\beta\rangle_3 + i\,|\beta\rangle_1|\chi\rangle_2|\alpha\rangle_3) \qquad (19.15)$$

$$|\psi\rangle_{II} = \frac{1}{\sqrt{3}}(|\alpha\rangle_1|\beta\rangle_2|\chi\rangle_3 + |\chi\rangle_1|\alpha\rangle_2|\beta\rangle_3 - i\,|\beta\rangle_1|\chi\rangle_2|\alpha\rangle_3) \qquad (19.16)$$

$$|\psi\rangle_I = \frac{1}{\sqrt{3}}(i\,|\alpha\rangle_1|\beta\rangle_2|\chi\rangle_3 + j\,|\chi\rangle_1|\alpha\rangle_2|\beta\rangle_3 + k\,|\beta\rangle_1|\chi\rangle_2|\alpha\rangle_3). \qquad (19.17)$$

19.3 The quantum entanglement probability amplitude for $n = N = 6$

The experimental diagram applicable to the $n = N = 6$ configuration for quantum entanglement is depicted in figure 19.2. Here, s is the photon source of three quanta pairs emitted in three directions at a relative angle of $\pi/3$.

Again, the first essential assumption is that the quanta emitted by the source are indistinguishable. Then, assuming that all detectors are identical, applying the Dirac interferometric principle, and following the methodology taught in the previous section,

$$|s\rangle = (|D\rangle_6 + |D\rangle_5 + |D\rangle_4 + |D\rangle_3 + |D\rangle_2 + |D\rangle_1). \qquad (19.18)$$

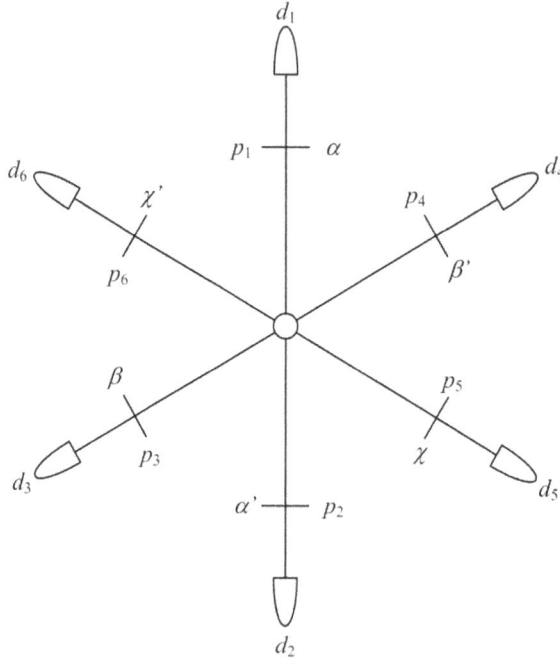

Figure 19.2. Experimental diagram applicable to quantum entanglement for the $n = N = 6$ configuration.

Then, in reference to figure 19.2 depicting the polarization alternatives (α, α'), (β, β'), and (χ, χ'), the various combined states are given by

$$|C\rangle_1 = |\alpha\rangle_1 |\alpha'\rangle_2 |\beta\rangle_3 |\beta'\rangle_4 |\chi\rangle_5 |\chi'\rangle_6 \tag{19.19}$$

$$|C\rangle_2 = |\alpha'\rangle_1 |\alpha\rangle_2 |\beta'\rangle_3 |\beta\rangle_4 |\chi'\rangle_5 |\chi\rangle_6 \tag{19.20}$$

$$|C\rangle_3 = |\chi\rangle_1 |\chi'\rangle_2 |\alpha\rangle_3 |\alpha'\rangle_4 |\beta\rangle_5 |\beta'\rangle_6 \tag{19.21}$$

$$|C\rangle_4 = |\chi'\rangle_1 |\chi\rangle_2 |\alpha'\rangle_3 |\alpha\rangle_4 |\beta'\rangle_5 |\beta\rangle_6 \tag{19.22}$$

$$|C\rangle_5 = |\beta\rangle_1 |\beta'\rangle_2 |\chi\rangle_3 |\chi'\rangle_4 |\alpha\rangle_5 |\alpha'\rangle_6 \tag{19.23}$$

$$|C\rangle_6 = |\beta'\rangle_1 |\beta\rangle_2 |\chi'\rangle_3 |\chi\rangle_4 |\alpha'\rangle_5 |\alpha\rangle_6. \tag{19.24}$$

Next, following the format of

$$|\psi\rangle_R = \frac{1}{\sqrt{N}} \sum_{j=1}^{N} (\pm) q |C\rangle_{N+1-j} \tag{19.25}$$

where q represents either 1 or a quaternion i, j, or k needed to satisfy the normalization condition,

$$1 = \||\psi\rangle_I\|^2 + \||\psi\rangle_{II}\|^2 + \||\psi\rangle_{III}\|^2 + \||\psi\rangle_{IV}\|^2 + \||\psi\rangle_V\|^2 + \||\psi\rangle_{VI}\|^2. \tag{19.26}$$

Thus, the individual probability amplitudes become

$$|\psi\rangle_I = \frac{1}{\sqrt{6}}(|C\rangle_1 + |C\rangle_2 + |C\rangle_3 + |C\rangle_4 + i\,|C\rangle_5 + i\,|C\rangle_6) \tag{19.27}$$

$$|\psi\rangle_{II} = \frac{1}{\sqrt{6}}(|C\rangle_1 + |C\rangle_2 + |C\rangle_3 + |C\rangle_4 - i\,|C\rangle_5 - i\,|C\rangle_6) \tag{19.28}$$

$$|\psi\rangle_{III} = \frac{1}{\sqrt{6}}(|C\rangle_1 + |C\rangle_2 - |C\rangle_3 - |C\rangle_4 + i\,|C\rangle_5 + i\,|C\rangle_6) \tag{19.29}$$

$$|\psi\rangle_{IV} = \frac{1}{\sqrt{6}}(|C\rangle_1 + |C\rangle_2 - |C\rangle_3 - |C\rangle_4 - i\,|C\rangle_5 - i\,|C\rangle_6) \tag{19.30}$$

$$|\psi\rangle_V = \frac{1}{\sqrt{6}}(i\,|C\rangle_1 + i\,|C\rangle_2 + j\,|C\rangle_3 + j\,|C\rangle_4 + k\,|C\rangle_5 + k\,|C\rangle_6) \tag{19.31}$$

$$|\psi\rangle_{VI} = \frac{1}{\sqrt{6}}(-i\,|C\rangle_1 - i\,|C\rangle_2 - j\,|C\rangle_3 - j\,|C\rangle_4 - k\,|C\rangle_5 - k\,|C\rangle_6). \tag{19.32}$$

Substitution of the states (19.19)–(19.24) into equations (19.27)–(19.32) yield the explicit probability amplitudes as a function of (α, α'), (β, β'), and (χ, χ').

19.4 Discussion

Examination of quantum entanglement involving n quanta and N propagation channels in $n = N = 3, 6$ configurations leads to the following conclusions:

1. Each individual probability amplitude includes $n = N$ terms.
2. The normalization factor for a quantum entanglement configuration including N propagation channels is $1/\sqrt{N}$.
3. In order to satisfy the normalization condition, the use of Hamilton's quaternions is necessary.
4. For a given $n = N$ entanglement configuration, there are $n = N$ probability amplitudes.

By observing equations (19.12)–(19.14) and (19.27)–(19.32), the equations applicable to $n = N = 9$ can be inferred.

References

Dirac P A M 1978 *The Principles of Quantum Mechanics* 4th edn (Oxford: Oxford University Press)

Duarte F J 2015 *Tunable Laser Optics* 2nd edn (New York: CRC)

Duarte F J 2016 Secure space-to-space interferometric communications and its nexus to the physics of quantum entanglement *Appl. Phys. Rev.* **3** 041301

Hamilton W R 1866 *Elements of Quaternions* (London: Longman Green & Co)

IOP Publishing

Fundamentals of Quantum Entanglement

F J Duarte

Chapter 20

What happens with the entanglement at $n = 1$ and $N = 2$?

It is observed that the derivational path between the Dirac–Feynman interferometric principle and the probability amplitude for quantum entanglement, $|\psi\rangle = (|x\rangle_1 |y\rangle_2 - |y\rangle_1 |x\rangle_2)$, is completely reversible. This observation has implications in regard to the foundations of quantum entanglement. Furthermore, the question of N propagation paths for $n = 1$ is examined.

20.1 Introduction

So far the quantum entanglement probability amplitudes applicable to $n = N = 2^1, 2^2, 2^3, 2^4 \ldots 2^r$ configurations have been examined. In this chapter the entanglement situation is considered at $n = 1$ and $N = 2^1$.

20.2 Reversibility: from entanglement to interference

We start from the now familiar probability amplitudes for quantum entanglement

$$|\psi\rangle_+ = \frac{1}{\sqrt{2}}(|x\rangle_1 |y\rangle_2 + |y\rangle_1 |x\rangle_2) \tag{20.1}$$

and its linear combination

$$|\psi\rangle_- = \frac{1}{\sqrt{2}}(|x\rangle_1 |y\rangle_2 - |y\rangle_1 |x\rangle_2). \tag{20.2}$$

The first thing to recognize is that these probability amplitudes are composed by two distinct vector states, and by applying the Dirac identity (Dirac 1978)

$$|X\rangle = |a\rangle_1 |b\rangle_2 |c\rangle_3 \ldots |g\rangle_n \tag{20.3}$$

in reverse

$$|x\rangle_1 |y\rangle_2 = |C\rangle_1 \qquad (20.4)$$

and

$$|x\rangle_2 |y\rangle_1 = |C\rangle_2 \qquad (20.5)$$

the first linear combination can be re-expressed as

$$|s\rangle = \frac{1}{\sqrt{2}}(|C\rangle_1 + |C\rangle_2) \qquad (20.6)$$

while its de-normalized version is

$$|s\rangle = (|C\rangle_1 + |C\rangle_2). \qquad (20.7)$$

Next, using the Dirac identity $|s\rangle = |j\rangle\langle j|s\rangle$ the states in equation (20.7) can be expressed as

$$|s\rangle_1 = |1\rangle\langle 1|s\rangle = |C\rangle_1 \qquad (20.8)$$

and

$$|s\rangle_2 = |2\rangle\langle 2|s\rangle = |C\rangle_2 \qquad (20.9)$$

so that equation (20.7), in its complete form, becomes

$$\langle x|s\rangle = (\langle x|2\rangle\langle 2|s\rangle + \langle x|1\rangle\langle 1|s\rangle) \qquad (20.10)$$

which is the probability amplitude for double-slit interference (Feynman *et al* 1965). As seen previously, equation (20.10) is the Dirac–Feynman principle

$$\langle x|s\rangle = \sum_{j=1}^{N}\langle x|j\rangle \langle j|s\rangle \qquad (20.11)$$

for one quantum, that is, $n = 1$, and two slits, that is, $N = 2$.

The process described here, from equation (20.1) to equation (20.11), is the derivation of the probability amplitude for quantum entanglement (Duarte 2013a, 2013b, 2014) *in reverse*.

20.3 Schematics

The schematics applicable to $n = N = 2^1$ quantum entanglement is illustrated in figure 20.1(a). As mentioned previously, the crucial assumptions in the derivation of equations (20.1) and (20.2) are as follows:

1. The quanta emitted by the source are indistinguishable, that is, $\nu_1 = \nu_2 = \nu$.
2. The detectors are assumed to be identical so that $d_1 = d_2 = d$.

These two characteristics are incorporated in the optical equivalent diagram depicted in figure 20.1(b) where the two mirrors are assumed to be 100% reflective non-polarizing surfaces. In essence, this is a dual beam path of indistinguishable

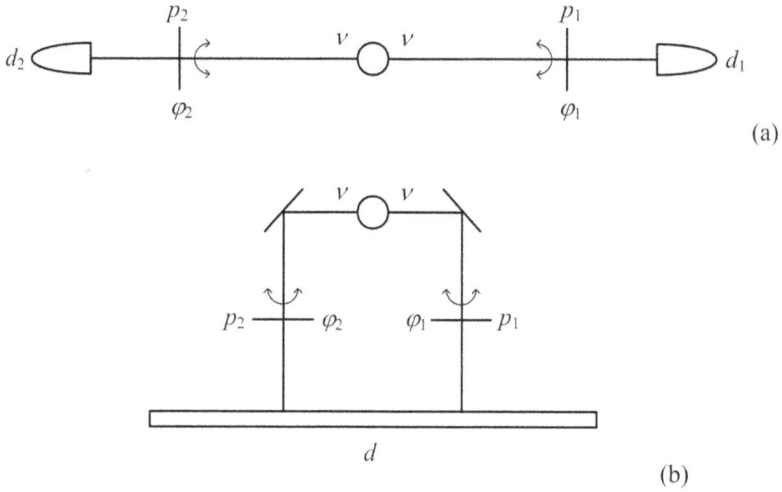

Figure 20.1. (a) Simplified experimental diagram applicable to quantum entanglement. Here, s is the photon source of indistinguishable quanta ($\nu_1 = \nu_2 = \nu$) emitted in the $+z$ and $-z$ directions. The angles φ_1 and φ_2 are the angles at the polarization analyzers measured on a plane perpendicular to the propagation axis while d_1 and d_2 are the corresponding detectors. (b) Optical equivalent diagram emphasizing the use of matched detectors $d_1 = d_2 = d$.

quanta propagating toward the detector via two polarizer analyzers set at angles φ_1 and φ_2 while their entangled behavior is described by

$$|\psi\rangle = \frac{1}{\sqrt{2}}(|x\rangle_1\,|y\rangle_2 \pm |y\rangle_1\,|x\rangle_2). \qquad (20.12)$$

The comparison is carried further in figure 20.2, where the equivalent entanglement diagram, applicable to equations (20.1) and (20.2), is shown next to a double-slit experiment. In figure 20.2(a), indistinguishable quanta, which are quantum mechanically the same quanta, propagate toward the detector via two polarizer analyzers set at angles φ_1 and φ_2. In figure 20.2(b), a single photon, or a population of indistinguishable photons, is spatially expanded (Duarte 2015) to illuminate slit 1 and slit 2 simultaneously; they lead to interference via

$$\langle x|s\rangle = (\langle x|2\rangle\langle 2|s\rangle + \langle x|1\rangle\langle 1|s\rangle).$$

20.4 Experimental and theoretical perspectives

In chapter 17 it was shown how to transition from the Dirac–Feynman interferometric principle to the quantum entanglement probability amplitude or

$$\langle x|s\rangle = \sum_{j=1}^{N}\langle x|j\rangle\,\langle j|s\rangle \rightarrow |\psi\rangle = \frac{1}{\sqrt{2}}(|x\rangle_1\,|y\rangle_2 \pm |y\rangle_1\,|x\rangle_2). \qquad (20.13)$$

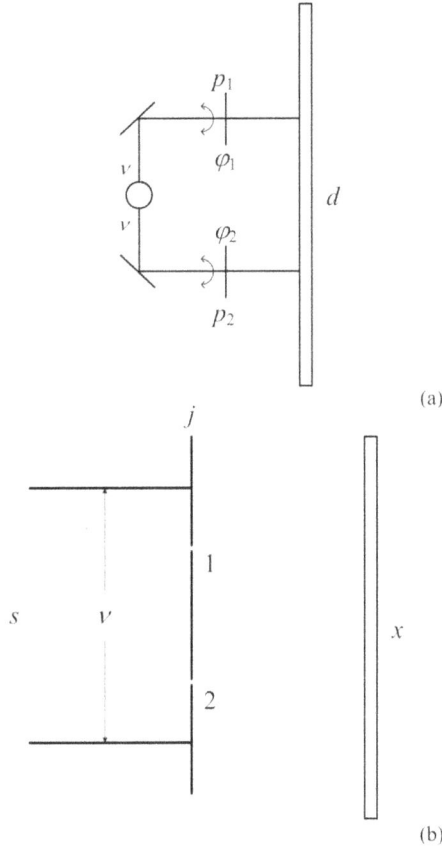

Figure 20.2. (a) Quantum entanglement optical equivalent diagram emphasizing indistinguishable quanta $(\nu_1 = \nu_2 = \nu)$ and the use of matched detectors $d_1 = d_2 = d$. (b) Double-slit interference diagram illustrating illumination by a single photon or a population of indistinguishable photons via an expanded beam profile.

This would seem to indicate that quantum interference is more fundamental than quantum entanglement. However, in the previous section it has been shown that the process is completely reversible:

$$|\psi\rangle = \frac{1}{\sqrt{2}}(|x\rangle_1 |y\rangle_2 \pm |y\rangle_1 |x\rangle_2) \rightarrow \langle x|s\rangle = \sum_{j=1}^{N}\langle x|j\rangle \langle j|s\rangle. \qquad (20.14)$$

The derivational path (20.13) indicates that quantum interference is the essence of quantum entanglement while the derivational path (20.14) indicates that quantum entanglement is the essence of quantum interference.

20.4.1 Experimental perspective

There is a rich history of investigating interference from the experimental perspective. The phenomenon of the interference of light has been documented since at least the early 1800s (Young 1802). Besides, interference as an experimental phenomenon is not only confined to optics but is also present in acoustics and hydraulics. Hence, there is an innate bias to favor interference as being the fundamental phenomenon that gives rise to quantum entanglement. If the Dirac–Feynman principle

$$\langle x|s \rangle = \sum_{j=1}^{N} \langle x|j \rangle \langle j|s \rangle$$

is considered as the fundamental principle, then

$$\langle x|s \rangle = \sum_{j=1}^{N} \langle x|j \rangle \langle j|s \rangle \rightarrow |\psi\rangle = \frac{1}{\sqrt{2}}(|x\rangle_1 |y\rangle_2 \pm |y\rangle_1 |x\rangle_2)$$

would disclose the hand of Nature.

Referring back to figure 20.2(b), the experimental situation described by

$$|s\rangle = (|2\rangle\langle 2|s\rangle + |1\rangle\langle 1|s\rangle) \tag{20.15}$$

suggests that the same photon is in two different states, a situation that can also be described as (Dirac 1978)

$$|a\rangle_1 |b\rangle_1. \tag{20.16}$$

These two different states of the same photon can be thought of as not just being entangled but being *utterly entangled*. This situation can arise in an ensemble of indistinguishable quanta where the quanta can be in different states. This subject is discussed further in chapter 28.

20.4.2 Theoretical perspective

Provided that

$$|\psi\rangle = \frac{1}{\sqrt{2}}(|x\rangle_1 |y\rangle_2 \pm |y\rangle_1 |x\rangle_2)$$

is derived in a rigorous and independent path (see chapter 24), then

$$|\psi\rangle = \frac{1}{\sqrt{2}}(|x\rangle_1 |y\rangle_2 \pm |y\rangle_1 |x\rangle_2) \rightarrow \langle x|s \rangle = \sum_{j=1}^{N} \langle x|j \rangle \langle j|s \rangle$$

is acceptable. A remarkable quote on this subject is worth mentioning at this stage: '$|x, y\rangle - |y, x\rangle$... was my first lesson in quantum mechanics, and in a very real sense my last, since the rest is mere technique, which can be learnt from books' (Ward 2004).

From a broader perspective, however, two further observations introduce further stimuli for thought:

1. Ward's comment following the Pryce–Ward publication in 1947: 'Wheeler proceeded to calculate ... But through the neglect of interference terms he derived an incorrect ... far too small value for the angular correlations ...' (Ward 1949).

2. Feynman's assertion in his lectures that interference is 'in the heart of quantum mechanics' (Feynman *et al* 1965).

20.5 Interference for N slits and $n = 1$

The generalized interferometric equation

$$|\langle x|s\rangle|^2 = \sum_{j=1}^{N} \Psi(r_j) \sum_{m=1}^{N} \Psi(r_m) e^{i(\Omega_m - \Omega_j)} \qquad (20.17)$$

originates in the Diracian probability amplitude

$$\langle x|s\rangle = \sum_{j=1}^{N} \langle x|j\rangle \langle j|s\rangle$$

which for *one photon illumination* means that all the individual probability amplitudes being generated at the j plane, that is, $\langle x|N\rangle$, $\langle x|N-1\rangle$... $\langle x|j\rangle$... $\langle x|3\rangle$, $\langle x|2\rangle$, $\langle x|1\rangle$, are *all intrinsically entangled* (Duarte and Taylor 2019).

References

Dirac P A M 1978 *The Principles of Quantum Mechanics* 4th edn (Oxford: Oxford University Press)

Duarte F J 2013a The probability amplitude for entangled polarizations: an interferometric approach *J. Mod. Opt.* **60** 1585–7

Duarte F J 2013b Tunable laser optics: applications to optics and quantum optics *Prog. Quantum Electron.* **37** 326–47

Duarte F J 2014 *Quantum Optics for Engineers* (New York: CRC)

Duarte F J 2015 *Tunable Laser Optics* 2nd edn (New York: CRC)

Duarte F J and Taylor T S 2019 *Unpublished*

Feynman R P, Leighton R B and Sands M 1965 *The Feynman Lectures on Physics* vol III (Reading: Addison-Wesley)

Ward J C 1949 *Some Properties of the Elementary Particles* (Oxford: Oxford University Press)

Ward J C 2004 *Memoirs of a Theoretical Physicist* (Rochester: Optics Journal)

Young T 1802 On the theory of light and colours *Philos. Trans. R. Soc. London* **92** 12–48

IOP Publishing

Fundamentals of Quantum Entanglement

F J Duarte

Chapter 21

Quantum entanglement probability amplitudes and Bell's theorem

Aspects of quantum polarization relevant to the probability amplitude for quantum entanglement, $|\psi\rangle = (|x\rangle_1|y\rangle_2 - |y\rangle_1|x\rangle_2)$, are reviewed. This enables the expression of this probability amplitude on explicit angular quantities that are then used to yield probabilities via Born's rule. These probabilities, in turn, are used to calculate Bell's inequalities.

21.1 Introduction

A concept that has prevailed over the fear of possible local hidden variable theories is Bell's theorem. This is the theorem, in a contemporaneous context, that proved that local hidden variable theories were incompatible with quantum mechanics.

Since Bell's theorem or Bell's inequalities are widely applied in the field of quantum cryptography, in this chapter a further exposition of the probability amplitudes for quantum entanglement in the context of Bell's theorem is given.

21.2 Probability amplitudes

For the case of $n = N = 2^1$ the probability amplitudes are

$$|\psi\rangle_I = \frac{1}{\sqrt{2}}(|x\rangle_1|y\rangle_2 + |y\rangle_1|x\rangle_2) \tag{21.1}$$

$$|\psi\rangle_{II} = \frac{1}{\sqrt{2}}(|x\rangle_1|y\rangle_2 - |y\rangle_1|x\rangle_2) \tag{21.2}$$

and for the case of $n = N = 2^2$ (Duarte 2015)

$$|\psi\rangle_I = \frac{1}{\sqrt{4}}(|x\rangle_1|y\rangle_2|\varphi\rangle_3|\varphi'\rangle_4 + |y\rangle_1|x\rangle_2|\varphi'\rangle_3|\varphi\rangle_4$$
$$+ |\varphi\rangle_1|\varphi'\rangle_2|x\rangle_3|y\rangle_4 + |\varphi'\rangle_1|\varphi\rangle_2|y\rangle_3|x\rangle_4) \tag{21.3}$$

$$|\psi\rangle_{II} = \frac{1}{\sqrt{4}}(|x\rangle_1|y\rangle_2|\varphi\rangle_3|\varphi'\rangle_4 + |y\rangle_1|x\rangle_2|\varphi'\rangle_3|\varphi\rangle_4$$
$$- |\varphi\rangle_1|\varphi'\rangle_2|x\rangle_3|y\rangle_4 - |\varphi'\rangle_1|\varphi\rangle_2|y\rangle_3|x\rangle_4)$$
(21.4)

$$|\psi\rangle_{III} = \frac{1}{\sqrt{4}}(|x\rangle_1|y\rangle_2|\varphi\rangle_3|\varphi'\rangle_4 - |y\rangle_1|x\rangle_2|\varphi'\rangle_3|\varphi\rangle_4$$
$$+ |\varphi\rangle_1|\varphi'\rangle_2|x\rangle_3|y\rangle_4 - |\varphi'\rangle_1|\varphi\rangle_2|y\rangle_3|x\rangle_4)$$
(21.5)

$$|\psi\rangle_{IV} = \frac{1}{\sqrt{4}}(|x\rangle_1|y\rangle_2|\varphi\rangle_3|\varphi'\rangle_4 - |y\rangle_1|x\rangle_2|\varphi'\rangle_3|\varphi\rangle_4$$
$$- |\varphi\rangle_1|\varphi'\rangle_2|x\rangle_3|y\rangle_4 + |\varphi'\rangle_1|\varphi\rangle_2|y\rangle_3|x\rangle_4).$$
(21.6)

Probability amplitudes for $n = N = 2^1, 2^2, 2^3 \ldots 2^r$ are given in chapter 18 and for $n = N = 3, 6$ in chapter 19.

Probability amplitudes are mathematical entities essential for the accurate representation of the physics at hand. Any modification to the physics, regardless of how insignificant it might seem, must be represented in the probability amplitude. However, probability amplitudes are not 'measurables'. The probability amplitude must be multiplied with its complex conjugate, according to the Born rule (Born 1926), to yield a probability that then becomes measurable. That is, for instance, using the probability amplitude given in equation (21.1),

$$P(\varphi_1, \varphi_2) = |\psi\rangle_I |\psi\rangle_I^*.$$
(21.7)

21.3 Quantum polarization

At this stage it is useful to briefly review some concepts of quantum polarization as introduced by Feynman *et al* (1965) and presented by Duarte (2014).

In figure 21.1, a system of axes illustrates axis rotation from x to x' and from y to y', that is, $x \rightarrow x'$ and $y \rightarrow y'$. From the geometry at hand,

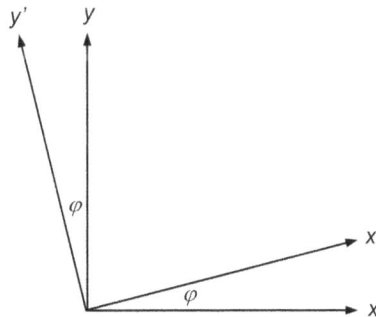

Figure 21.1. Set of x and y axes and set of rotating x' and y' axes.

$$\frac{x}{x'} = \cos \varphi \qquad (21.8)$$

$$\frac{x}{y'} = \sin \varphi \qquad (21.9)$$

$$\frac{y}{x'} = -\sin \varphi \qquad (21.10)$$

$$\frac{y}{y'} = \cos \varphi. \qquad (21.11)$$

Using the simple geometrical rotations given above,

$$\langle x|x' \rangle = \cos \varphi \qquad (21.12)$$

$$\langle y|x' \rangle = \sin \varphi \qquad (21.13)$$

$$\langle x|y' \rangle = -\sin \varphi \qquad (21.14)$$

$$\langle y|y' \rangle = \cos \varphi \qquad (21.15)$$

and it can also be shown that

$$\langle y|x' \rangle = \langle y'|x \rangle = \sin \varphi \qquad (21.16)$$

$$\langle x|y' \rangle = \langle x'|y \rangle = -\sin \varphi. \qquad (21.17)$$

Probability amplitudes of interest are

$$\langle x|x' \rangle = \langle x|x \rangle \langle x|x' \rangle + \langle x|y \rangle \langle y|x' \rangle \qquad (21.18)$$

$$\langle y'|x \rangle = \langle y'|x' \rangle \langle x'|x \rangle + \langle y'|y' \rangle \langle y'|x \rangle \qquad (21.19)$$

$$\langle x'|y \rangle = \langle x'|y' \rangle \langle y'|y \rangle + \langle x'|x' \rangle \langle x'|y \rangle \qquad (21.20)$$

$$\langle y|y' \rangle = \langle y|y \rangle \langle y|y' \rangle + \langle y|x \rangle \langle x|y' \rangle. \qquad (21.21)$$

Next, using equations (21.12)–(21.15) leads us to

$$\langle x|x' \rangle = \langle x|x \rangle \cos \varphi + \langle x|y \rangle \sin \varphi \qquad (21.22)$$

$$\langle y'|x \rangle = \langle y'|x' \rangle \cos \varphi + \langle y'|y' \rangle \sin \varphi \qquad (21.23)$$

$$\langle x'|y \rangle = \langle x'|y' \rangle \cos \varphi + \langle x'|x' \rangle (-\sin \varphi) \qquad (21.24)$$

$$\langle y|y' \rangle = \langle y|y \rangle \cos \varphi + \langle y|x \rangle (-\sin \varphi). \qquad (21.25)$$

Equations (21.22)–(21.25) in abstract form become (Duarte 2014)

$$|x' \rangle = |x \rangle \cos \varphi + |y \rangle \sin \varphi \qquad (21.26)$$

$$|x\rangle = |x'\rangle \cos\varphi + |y'\rangle \sin\varphi \tag{21.27}$$

$$|y\rangle = |y'\rangle \cos\varphi - |x'\rangle \sin\varphi \tag{21.28}$$

$$|y'\rangle = |y\rangle \cos\varphi - |x\rangle \sin\varphi \tag{21.29}$$

and equation (21.26) is the equation given by Feynman *et al* (1965).

Since $\langle x|x\rangle = 1$, $\langle y|y\rangle = 1$, and $\langle x|y\rangle = 0$, $\langle y|x\rangle = 0$, then

$$\langle x|x'\rangle = \cos\varphi \tag{21.30}$$

$$\langle y'|x\rangle = \sin\varphi \tag{21.31}$$

$$\langle x'|y\rangle = -\sin\varphi \tag{21.32}$$

$$\langle y|y'\rangle = \cos\varphi \tag{21.33}$$

and the probabilities of interest are evaluated as

$$|\langle x|x'\rangle|^2 = \cos^2\varphi \tag{21.34}$$

$$|\langle y'|x\rangle|^2 = \sin^2\varphi \tag{21.35}$$

$$|\langle x'|y\rangle|^2 = \sin^2\varphi \tag{21.36}$$

$$|\langle y|y'\rangle|^2 = \cos^2\varphi. \tag{21.37}$$

21.4 Quantum probabilities and Bell's theorem

By going back to the Pryce–Ward probability amplitude

$$|\psi\rangle = \frac{1}{\sqrt{2}}(|x\rangle_1|y\rangle_2 - |y\rangle_1|x\rangle_2) \tag{21.38}$$

and then completing the *bra–ket* expressions on the rhs, for notational convenience,

$$|\psi\rangle = \frac{1}{\sqrt{2}}\big(\langle x'|x\rangle_1\langle x'|y\rangle_2 - \langle x'|y\rangle_1\langle x'|x\rangle_2\big) \tag{21.39}$$

the probability becomes (Duarte 2014)

$$\||\psi\rangle|^2 = \frac{1}{2}\big(\langle x'|x\rangle_1\langle x'|y\rangle_2 - \langle x'|y\rangle_1\langle x'|x\rangle_2\big)^2. \tag{21.40}$$

Substitution of the corresponding individual probability amplitudes into equation (21.40) leads to

$$\||\psi\rangle|^2 = \frac{1}{2}(-\cos\varphi_1 \sin\varphi_2 + \sin\varphi_1 \cos\varphi_2)^2. \tag{21.41}$$

Using the appropriate geometrical identity (see appendix G), this probability can be expressed as

$$\||\psi\rangle|^2 = \frac{1}{2}\sin^2(\varphi_1 - \varphi_2). \tag{21.42}$$

By going back to the probability amplitude for quantum entanglement given in equation (21.39) and considering the situation in which quanta 1 is incident on the polarization analyzer 1 (p_1), as illustrated in figure 21.2, and quanta 2 is incident on the polarization analyzer 2 (p_2), then

$$|\psi\rangle = \frac{1}{\sqrt{2}}(\sin \varphi_1 \cos \varphi_2 - \cos \varphi_1 \sin \varphi_2) \tag{21.43}$$

$$|\psi\rangle = \frac{1}{\sqrt{2}}\sin(\varphi_1 - \varphi_2). \tag{21.44}$$

Moreover, if the polarization analyzers are rotated by $\pi/2$,

$$|\psi\rangle = \frac{1}{\sqrt{2}}\cos(\varphi_1 - \varphi_2). \tag{21.45}$$

Next, using the notation of Mandel and Wolf (1995), if transmission through a polarization analyzer is denoted by + and absorption by −, then the corresponding probabilities can be written as

$$p(+\varphi_1, +\varphi_2) = p(-\varphi_1, -\varphi_2) = \frac{1}{2}\sin^2(\varphi_1 - \varphi_2) \tag{21.46}$$

$$p(+\varphi_1, -\varphi_2) = p(-\varphi_1, +\varphi_2) = \frac{1}{2}\cos^2(\varphi_1 - \varphi_2). \tag{21.47}$$

The overall probability $P(\varphi_1, \varphi_2)$ is the sum of the individual probability alternatives (Mandel and Wolf 1995)

$$P(\varphi_1, \varphi_2) = p(+\varphi_1, +\varphi_2) + p(-\varphi_1, -\varphi_2) - p(+\varphi_1, -\varphi_2) - p(-\varphi_1, +\varphi_2) \tag{21.48}$$

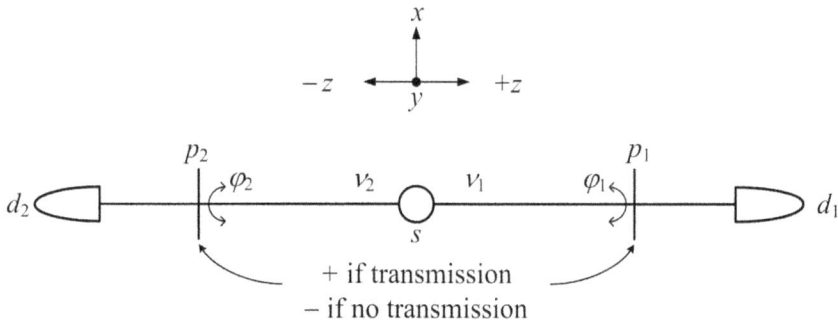

Figure 21.2. Quantum entanglement diagram appropriate for probability calculations.

$$P(\varphi_1, \varphi_2) = \sin^2(\varphi_1 - \varphi_2) - \cos^2(\varphi_1 - \varphi_2)$$

and by using the geometrical identities given in appendix G the overall probability can be expressed as

$$P(\varphi_1, \varphi_2) = -\cos 2(\varphi_1 - \varphi_2). \tag{21.49}$$

The reader should observe that here it is again: the Pryce–Ward polarization differential angle first introduced in chapter 6

$$\Delta\varphi = 2(\varphi_1 - \varphi_2). \tag{21.50}$$

As seen in chapter 10, Bell's inequality, also known as Bell's theorem, can be stated as (Bell 1964)

$$|P(x, y) - P(x, y')| + |P(x', y') + P(x', y)| \leqslant 2 \tag{21.51}$$

and using the angular notation becomes

$$|P(\varphi_1, \varphi_2) - P(\varphi_1, \varphi_2')| + |P(\varphi_1', \varphi_2') + P(\varphi_1', \varphi_2)| \leqslant 2 \tag{21.52}$$

which, in shorthand notation, can be expressed as

$$\Sigma_P \leqslant 2 \tag{21.53}$$

where

$$\Sigma_P = |P(\varphi_1, \varphi_2) - P(\varphi_1, \varphi_2')| + |P(\varphi_1', \varphi_2') + P(\varphi_1', \varphi_2)|. \tag{21.54}$$

21.5 Example

As an example, consider $\varphi_1 = \pi/20$, $\varphi_2 = 9\pi/20$, $\varphi_1' = \pi/3$, and $\varphi_2' = \pi/6$:

$$P(\varphi_1, \varphi_2) = -\cos 2(\varphi_1 - \varphi_2) \tag{21.55}$$

$$P(\varphi_1, \varphi_2') = -\cos 2(\varphi_1 - \varphi_2') \tag{21.56}$$

$$P(\varphi_1', \varphi_2') = -\cos 2(\varphi_1' - \varphi_2') \tag{21.57}$$

$$P(\varphi_1', \varphi_2) = -\cos 2(\varphi_1' - \varphi_2) \tag{21.58}$$

for which the left-hand side of Bell's inequality becomes

$$|P(\varphi_1, \varphi_2) - P(\varphi_1, \varphi_2')| + |P(\varphi_1', \varphi_2') + P(\varphi_1', \varphi_2)| = |+0.81 + 0.74| \\ + |-0.50 - 0.74| \tag{21.59}$$

or

$$\Sigma_P = |+0.81 + 0.74| + |-0.50 - 0.74| \tag{21.60}$$

$$\Sigma_P = 2.79$$

and Bell's inequality is violated. For the sake of completeness, expressing the overall numerical equality leads to

$$|+0.81 + 0.74| + |-0.5 - 0.74| \leqslant 2 \qquad (21.61)$$

$$2.79 \leqslant 2 \qquad (21.62)$$

which is obviously *not true*, thus demonstrating that Bell's inequality is violated when using the quantum probabilities computed via the Pryce–Ward probability amplitude (Pryce and Ward 1947, Ward 1949) for quantum entanglement. See chapter 10 for an additional example.

21.6 Discussion

Violation of Bell's inequalities reinforces the fact that hidden variable theories are incompatible with quantum mechanics. More specifically, it means that hidden variable theories are not compatible with probabilities calculated from the probability amplitudes of quantum entanglement. This subject is further discussed in chapters 28 and 29.

The results of quantum entanglement experiments are fully characterized by their corresponding probability amplitudes and measurable probabilities. However, in the way and manner that this subject has evolved, Bell's theorem and the corresponding Bell's inequalities have become widely used to ensure that the quantum mechanical mechanics have not been biased by possible third-party intrusion.

References

Bell J S 1964 On the Einstein–Podolsky–Rosen paradox *Physics* **1** 195–200

Born M 1926 Zur quantenmechanik der stoßvorgänge *Z. Phys.* **37** 863–7

Duarte F J 2014 *Quantum Optics for Engineering* (New York: CRC)

Duarte F J 2015 *Tunable Laser Optics* 2nd edn (New York: CRC)

Feynman R P, Leighton R B and Sands M 1965 *The Feynman Lectures on Physics* vol III (Reading, MA: Addison-Wesley)

Mandel L and Wolf E 1995 *Optical Coherence and Quantum Optics* (Cambridge: Cambridge University Press)

Pryce M L H and Ward J C 1947 Angular correlation effects with annihilation radiation *Nature* **160** 435

Ward J C 1949 *Some Properties of the Elementary Particles* (Oxford: Oxford University Press)

IOP Publishing

Fundamentals of Quantum Entanglement

F J Duarte

Chapter 22

Cryptography via quantum entanglement

The application of the probability amplitude for quantum entanglement, $|\psi\rangle = (|x\rangle_1 |y\rangle_2 - |y\rangle_1 |x\rangle_2)$, to quantum cryptography is described.

22.1 Introduction

Cryptography is a Greek-based word that conveys the concepts of *hidden writing* or *secret writing*. Traditionally, cryptography has involved a *secret key* or *code* shared between two parties, the interchange of coded messages between the parties, and the subsequent decoding of the messages using the key or code. The weak link in this arrangement is the key or code that can be captured and deciphered or decrypted by a third party, spy, or eavesdropper.

Quantum cryptography is the application of the physics of quantum entanglement, and quantum technologies, to enable the secure creation, transmission, and reception of characters, words, or messages. According to Bennett (1992), quantum key distribution is 'the classically impossible task of distributing secret information over an insecure channel whose transmissions are subject to inspection by an eavesdropper'.

Quantum cryptography was first discussed, as a concept, by S Weisner in the 1970s. A scheme to implement quantum cryptography was subsequently introduced in the open literature by Bennett and Brassard (1984) in what became known as the Bennett–Brassard approach. Further developments were published by Bennett *et al* (1992a). The Bennett–Brassard approach utilizes the straightforward polarization properties of propagating quanta such as

$$|H\rangle = |x\rangle \tag{22.1}$$

$$|V\rangle = |y\rangle \tag{22.2}$$

doi:10.1088/2053-2563/ab2b33ch22

$$|L\rangle = \frac{1}{\sqrt{2}}(|x\rangle - i\,|y\rangle) \tag{22.3}$$

$$|R\rangle = \frac{1}{\sqrt{2}}(|x\rangle + i\,|y\rangle) \tag{22.4}$$

where $|L\rangle$ refers to left-circularly polarized photons, and $|R\rangle$ refers to right-circularly polarized photons. This methodology is not considered here and for a more detailed description the reader should refer to Bennett (1992), Bennett and Brassard (2014), Bennett *et al* (1992a, 1992b), or Duarte (2014).

The focus of this chapter is on free-space quantum cryptography derived from the physics of the quantum entanglement probability amplitude. For an extensive review on quantum cryptography, and the various approaches to implement it, the reader may refer to Gisin *et al* (2002).

22.2 Measurement protocol

Pryce–Ward-type probability amplitudes (Pryce and Ward 1947, Ward 1949) of the form (see chapter 17)

$$|\psi\rangle_+ = \frac{1}{\sqrt{2}}(|x\rangle_1\,|y\rangle_2 + |y\rangle_1\,|x\rangle_2) \tag{22.5}$$

$$|\psi\rangle_- = \frac{1}{\sqrt{2}}(|x\rangle_1\,|y\rangle_2 - |y\rangle_1\,|x\rangle_2) \tag{22.6}$$

$$|\psi\rangle^+ = \frac{1}{\sqrt{2}}(|x\rangle_1\,|x\rangle_2 + |y\rangle_1\,|y\rangle_2) \tag{22.7}$$

$$|\psi\rangle^- = \frac{1}{\sqrt{2}}(|x\rangle_1\,|x\rangle_2 - |y\rangle_1\,|y\rangle_2) \tag{22.8}$$

were *implicitly* introduced to the field of quantum cryptography by Ekert (1991), who described a quantum cryptographic methodology utilizing spin-1/2 particles and Bell's theorem (Bell 1964) via the Bohm–EPR perspective (Einstein *et al* 1935, Bohm 1951). The word 'implicitly' is used here because, although not mentioned by Ekert, as soon as Bell's inequalities are utilized that immediately requires the calculation of quantum probabilities, and the relevant quantum probabilities are calculated via the probability amplitudes given in equations (22.5)–(22.8).

The Ekert (1991) approach is based on the 'completeness of quantum mechanics'. In other words, it is the violation of Bell's inequalities that ensures security.

Bosonic rather than fermionic realizations of the Ekert approach were published by Naik *et al* (2000), Jennewein *et al* (2000), and Tittel *et al* (2000). These experimenters utilized their own version of the Ekert approach but the following features provide a fair summary.

First, one of the probability amplitudes for quantum entanglement is identified:

$$|\psi\rangle_- = \frac{1}{\sqrt{2}}(|H\rangle_1 |V\rangle_2 - |V\rangle_1 |H\rangle_2) \tag{22.9}$$

$$|\psi\rangle^+ = \frac{1}{\sqrt{2}}(|H\rangle_1 |H\rangle_2 + |V\rangle_1 |V\rangle_2) \tag{22.10}$$

where $|H\rangle = |x\rangle$ and $|V\rangle = |y\rangle$. Then each communicating party, A and B, proceeds to measure using bases such as (Naik *et al* 2000)

$$\left(|H\rangle_1 + e^{i\varphi_j^A} |V\rangle_1\right) \tag{22.11}$$

and

$$\left(|H\rangle_2 + e^{i\varphi_j^B} |V\rangle_2\right) \tag{22.12}$$

where the angular quantities φ_j^A and φ_j^B ($j = 1, 2, 3, 4...$) assume *random* values from various options such as, $\varphi_1^A = \pi/4$, $\varphi_2^A = \pi/2$, $\varphi_3^A = 3\pi/4$, and $\varphi_4^A = \pi$, and $\varphi_1^B = 0$, $\varphi_2^B = \pi/4$, $\varphi_3^B = \pi/2$, and $\varphi_4^B = 3\pi/4$ (Ekert 1991, Naik *et al* 2000).

Next, A and B disclose the bases used in the measurements but not the results of the measurements. For measurements involving $(\varphi_j^A + \varphi_j^B) = \pi$, the results are completely correlated and that becomes the key. The results from other combinations are made public and used to calculate the corresponding Bell's inequality. Violation of Bell's inequality ensures that the measurements are secure.

Two final comments are relevant: first, the quantum entanglement branch of quantum cryptography, which actually depends entirely on the Pryce–Ward probability amplitude, appears to have been conceived entirely within the philosophical route via Bell's theorem. Second, Bennett *et al* (1992b) indicate that the Ekert approach and the Bennet and Brassard approach are equivalent. These authors also argue that neither the EPR effect (Einstein *et al* 1935) nor Bell's inequalities (Bell 1964) are 'essential' for security certification.

An interesting historic note here is that probability amplitude equations of the form described by equations (22.5)–(22.10) became known as *Bell states* even though Bell was absent from their development.

22.3 Experiments

The bosonic experimental configuration necessary for quantum cryptography is, in principle, the same as that used for the testing of Bell's inequalities by Aspect *et al* (1982). A generic experimental configuration to this effect is illustrated in figure 22.1. A source of entangled photon pairs emits quanta 1 in the $+z$ direction, toward participant A, and quanta 2 in the $-z$ direction, toward participant B. In addition to the random emission of photons pairs with polarizations orthogonal to each other, the analyzers are also set randomly at angles φ_j^A and φ_j^B.

In practice the situation is not as neat as described above. First of all, the photon pair sources of choice, based on optically pumped type-II spontaneous parametric

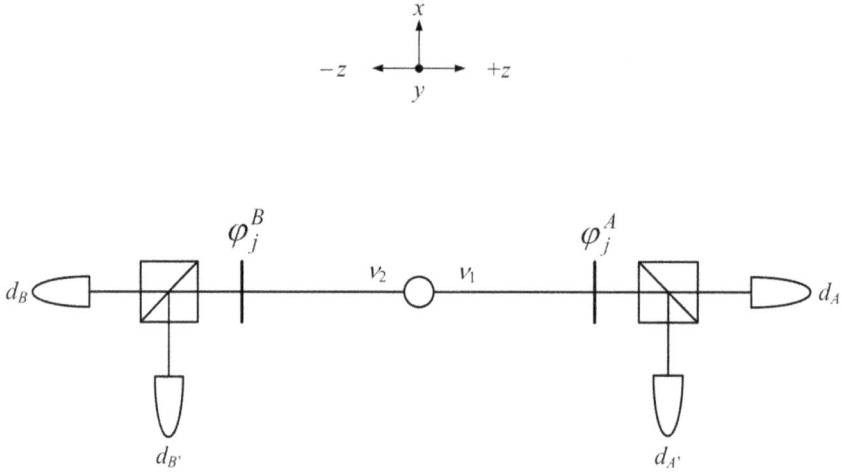

Figure 22.1. Schematics for an idealized quantum key distribution experiment. The settings of φ_j^A and φ_j^B are randomized (see text).

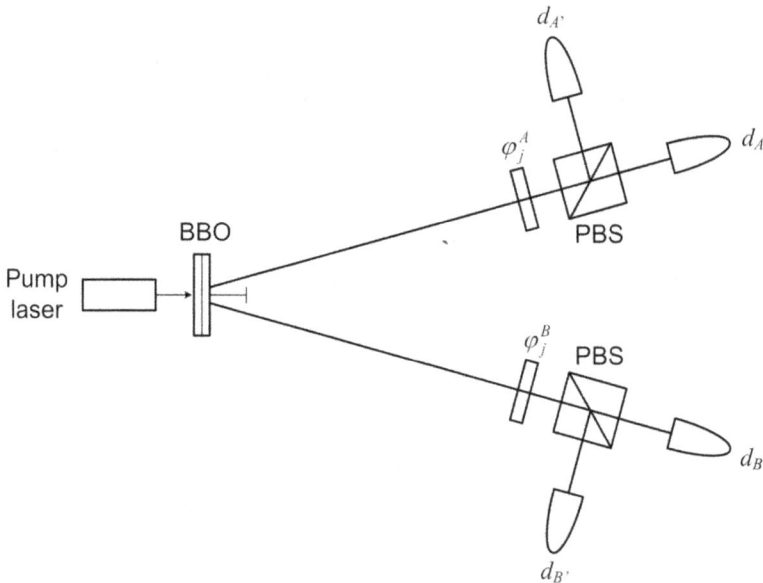

Figure 22.2. Generic schematics quantum key distribution experiment utilizing SPDC. In this case, an appropriate pump laser is used to excite a type-II crystal configuration often comprised of **BBO** crystals. Prior to the polarizing beam splitter (PBS), an appropriate polarizer analyzer capable of providing randomized settings for φ_j^A and φ_j^B is deployed. In some experiments the source of entangled photons is located near one of the communicating parties, such as A.

down-conversion (SPDC), can be rather inefficient and they emit the photon pairs not collinearly in opposite directions but at an angle relative to each other as illustrated in figure 22.2 and chapter 27. Furthermore, there are different and varied methods to achieve the random settings for φ_j^A and φ_j^B, and thus the experimental configurations from experiment to experiment tend to vary.

An experiment worth noting is that described by Ursin *et al* (2007) that was conducted in free space over a distance of over 144 km. These authors report a strong violation of Bell's inequality with $\Sigma_p = 2.508 \pm 0.037$ and the implementation of quantum key distribution.

References

Aspect A, Grangier P and Roger G 1982 Experimental test of Bell's inequality using time-varying analyzers *Phys. Rev. Lett.* **49** 1804–7

Bell J S 1964 On the Einstein–Podolsky–Rosen paradox *Physics* **1** 195–200

Bennett C H 1992 Quantum cryptography using any two nonorthogonal states *Phys. Rev. Lett.* **68** 3121–4

Bennett C H, Bessette F, Brassard G, Salvail L and Smolin J 1992a Experimental quantum cryptography *J. Cryptol.* **5** 3–28

Bennett C H, Brassard G and Mermin N D 1992b Quantum cryptography without Bells theorem *Phys. Rev. Lett.* **68** 557–9

Bennett C H and Brassard G 1984 Quantum cryptography: public key distribution and coin tossing *Proc. of IEEE Int. Conf. on Computers Systems and Signal Processing (Bangalore, India)*

Bennett C H and Brassard G 2014 Quantum cryptography: public key distribution and coin tossing *Theor. Comput. Sci.* **560** 7–11

Bohm D 1951 *Quantum Theory* (Englewood Cliffs, NJ: Prentice-Hall)

Duarte F J 2014 *Quantum Optics for Engineers* (New York: CRC)

Einstein A, Podolsky B and Rosen N 1935 Can quantum mechanical description of physical reality be considered complete? *Phys. Rev.* **47** 777–80

Ekert A K 1991 Quantum cryptography based on Bell's theorem *Phys. Rev. Lett.* **67** 661–3

Gisin N, Ribordy G, Tittel W and Zbinden H 2002 Quantum cryptography *Rev. Mod. Phys.* **74** 145–95

Jennewein T, Simon C, Weihs G, Weinfurter H and Zeilinger A 2000 Quantum cryptography with entangled photons *Phys. Rev. Lett.* **84** 4729–32

Naik D S, Peterson C G, White A G, Berglund A J and Kwiat P G 2000 Entangled state quantum cryptography: eavesdropping on the Ekert protocol *Phys. Rev. Lett.* **84** 4733–6

Pryce M H L and Ward J C 1947 Angular correlation effects with annihilation radiation *Nature* **160** 435

Tittel W, Brendel J, Zbinden H and Gisin N 2000 Quantum cryptography using entangled photons in energy-time Bell states *Phys. Rev. Lett.* **84** 4737–40

Ursin R *et al* 2007 Free-space distribution of entanglement and single photons over 144 km *Nat. Phys.* **3** 481–6

Ward J C 1949 *Some Properties of the Elementary Particles* (Oxford: Oxford University Press)

Chapter 23

Quantum entanglement and teleportation

The application of the probability amplitude for quantum entanglement, $|\psi\rangle = (|x\rangle_1|y\rangle_2 - |y\rangle_1|x\rangle_2)$, to quantum teleportation is described.

23.1 Introduction

Quantum teleportation, as described by Bennett *et al* (1993), consists of the disintegration of one quantum state, at the emitter's site, and the subsequent reintegration, or reassembling, of that quantum state at the receiver's site.

This teleportation of a quantum state, from one site to another site, is based on the physics made possible by the Pryce–Ward probability amplitude

$$|\psi\rangle = \frac{1}{\sqrt{2}}(|x\rangle_1|y\rangle_2 \pm |y\rangle_1|x\rangle_2). \qquad (23.1)$$

In the notation utilized in this chapter, the polarization $|x\rangle$ and $|y\rangle$ states are represented by $|1\rangle$ and $|0\rangle$, and their corresponding vectors (Fowles 1968) are

$$|x\rangle = |1\rangle = \begin{pmatrix} 1 \\ 0 \end{pmatrix} \qquad (23.2)$$

and

$$|y\rangle = |0\rangle = \begin{pmatrix} 0 \\ 1 \end{pmatrix}. \qquad (23.3)$$

The emphasis of this section on the application of the quantum entanglement probability amplitude is on principles. For reviews on the subject, including recent descriptions of the technology, the reader may refer to Pirandola *et al* (2015).

Note: as long as consistency is maintained, the transverse polarization states $|x\rangle$ and $|y\rangle$ can be defined as either $|1\rangle$ and $|0\rangle$, or $|0\rangle$ and $|1\rangle$. If they are defined as $|0\rangle$ and $|1\rangle$, then the corresponding equations of entanglement are $|\psi\rangle = 2^{-1/2}(|0\rangle_1|1\rangle_2 \pm |1\rangle_1|0\rangle_2)$

and $|\psi\rangle = 2^{-1/2}(|0\rangle_1|0\rangle_2 \pm |1\rangle_1|1\rangle_2)$. If, on the other hand, they are defined as $|1\rangle$ and $|0\rangle$, then the corresponding entanglement equations are $|\psi\rangle = 2^{-1/2}(|1\rangle_1|0\rangle_2 \pm |0\rangle_1|1\rangle_2)$ and $|\psi\rangle = 2^{-1/2}(|1\rangle_1|1\rangle_2 \pm |0\rangle_1|0\rangle_2)$. This is explained in detail in appendix F.

23.2 The mechanics of teleportation

In this section, the mechanics of quantum teleportation is described. That is, the mechanics of disassembling a quantum state at the emitter's site and the subsequent reassembling of the same quantum state at the receiver's site. This exposition uses the style and notation of a previous review on the subject (Duarte 2014). As a preamble, it is useful to introduce the 2×2 identity matrix I and Pauli's matrices (Feynman *et al* 1965)

$$I = \begin{pmatrix} 1 & 0 \\ 0 & 1 \end{pmatrix} \tag{23.4}$$

$$\sigma_x = \begin{pmatrix} 0 & 1 \\ 1 & 0 \end{pmatrix} \tag{23.5}$$

$$\sigma_y = \begin{pmatrix} 0 & -i \\ i & 0 \end{pmatrix} \tag{23.6}$$

$$\sigma_z = \begin{pmatrix} 1 & 0 \\ 0 & -1 \end{pmatrix}. \tag{23.7}$$

Next, for background information, it is necessary to reintroduce various forms of the Pryce–Ward probability amplitudes as

$$|s\rangle_+ = \frac{1}{\sqrt{2}}(|1\rangle_1|0\rangle_2 + |0\rangle_1|1\rangle_2) \tag{23.8}$$

$$|s\rangle_- = \frac{1}{\sqrt{2}}(|1\rangle_1|0\rangle_2 - |0\rangle_1|1\rangle_2) \tag{23.9}$$

$$|r\rangle_+ = \frac{1}{\sqrt{2}}(|1\rangle_1|1\rangle_2 + |0\rangle_1|0\rangle_2) \tag{23.10}$$

$$|r\rangle_- = \frac{1}{\sqrt{2}}(|1\rangle_1|1\rangle_2 - |0\rangle_1|0\rangle_2). \tag{23.11}$$

Adding and subtracting these equations yield the following combined states:

$$\frac{1}{\sqrt{2}}(|s\rangle_+ + |s\rangle_-) = |1\rangle_1|0\rangle_2 \tag{23.12}$$

$$\frac{1}{\sqrt{2}}(|s\rangle_+ - |s\rangle_-) = |0\rangle_1|1\rangle_2 \tag{23.13}$$

$$\frac{1}{\sqrt{2}}(|r\rangle_+ + |r\rangle_-) = |1\rangle_1|1\rangle_2 \tag{23.14}$$

$$\frac{1}{\sqrt{2}}(|r\rangle_+ - |r\rangle_-) = |0\rangle_1|0\rangle_2. \tag{23.15}$$

The mechanics of teleporting one quantum state from the transmitter's site to the receiver's site can be illustrated using the scheme disclosed by Kim *et al* (2001).

Assume that the following arbitrary quantum state is to be teleported:

$$|\phi\rangle_1 = (\alpha |0\rangle_1 + \beta |1\rangle_1) \tag{23.16}$$

where $|1\rangle$ and $|0\rangle$ are the two orthogonal vectors of polarization corresponding to $|x\rangle$ and $|y\rangle$, respectively. Factors α and β are related by $|\alpha|^2 + |\beta|^2 = 1$. This probability amplitude is the state to be disassembled at the emitter's site and that will be subsequently replicated at the receiver's site. First, the emitter generates an entangled state of the form of any of the probability amplitudes given in equations (23.8)–(23.11). Next, assume that the entangled state selected to perform the operation is that given in equation (23.8). With a slight change in notation this state can be re-expressed as

$$|s\rangle_{23+} = \frac{1}{\sqrt{2}}(|1\rangle_2|0\rangle_3 + |0\rangle_2|1\rangle_3). \tag{23.17}$$

Next, the emitter creates a three-particle state $|\varphi\rangle_{123}$ via the multiplication of $|\phi\rangle_1$ and $|s\rangle_{23+}$:

$$|\varphi\rangle_{123} = \frac{1}{\sqrt{2}}(\alpha |0\rangle_1|1\rangle_2|0\rangle_3 + \alpha |0\rangle_1|0\rangle_2|1\rangle_3 + \beta |1\rangle_1|1\rangle_2|0\rangle_3 + \beta |1\rangle_1|0\rangle_2|1\rangle_3). \tag{23.18}$$

This operation makes possible the projection of photon 3 on the input state given in equation (23.16). By using equations (23.12)–(23.15) while noting that $|s\rangle_+ = |s\rangle_{12+}$, $|s\rangle_- = |s\rangle_{12-}$, $|r\rangle_+ = |r\rangle_{12+}$, and $|r\rangle_- = |r\rangle_{12-}$, the three-particle state $|\varphi\rangle_{123}$, that is, equation (23.18), can be rewritten as

$$|\varphi\rangle_{123} = \frac{1}{2}(|s\rangle_{12+}|\phi\rangle_{3+} + |s\rangle_{12-}|\phi\rangle_{3-} + |r\rangle_{12+}|\vartheta\rangle_{3+} + |r\rangle_{12-}|\vartheta\rangle_{3-}) \tag{23.19}$$

meaning that each of the four component states has an equal 25% chance of materializing. In equation (23.19), photon 3 is projected into the following states:

$$|\phi\rangle_{3+} = (\alpha |0\rangle_3 + \beta |1\rangle_3) \tag{23.20}$$

$$|\phi\rangle_{3-} = (\beta |1\rangle_3 - \alpha |0\rangle_3) \tag{23.21}$$

$$|\vartheta\rangle_{3+} = (\alpha \, |1\rangle_3 + \beta \, |0\rangle_3) \qquad (23.22)$$

$$|\vartheta\rangle_{3-} = (\beta \, |0\rangle_3 - \alpha \, |1\rangle_3). \qquad (23.23)$$

Following the multiplication of $|\phi\rangle_1$ and $|s\rangle_{23+}$, the original state to be teleported has been disassembled and now photon 3 is projected on the original input state.

In this procedure, if the emitter measures $|s\rangle_{12+}$, the qubit communicated to the receiver is $|\phi\rangle_{3+}$. In this case, the only unitary transformation needed is just the identity matrix and the receiver's task is done. However, if $|s\rangle_{12-}$ is communicated, then the receiver must apply the appropriate transformation on $|\phi\rangle_{3-}$ to recover the original quantum state. The appropriate unitary transformation needed is communicated classically by the emitter to the receiver (Kim *et al* 2001).

This process can be stated in short hand as $|s\rangle_{12+} \rightarrow |\phi\rangle_{3+}$, where $\rightarrow$ stands for 'to be communicated to the receiver': The other alternatives are $|s\rangle_{12-} \rightarrow |\phi\rangle_{3-}$, $|r\rangle_{12+} \rightarrow |\vartheta\rangle_{3+}$, and $|r\rangle_{12-} \rightarrow |\vartheta\rangle_{3-}$.

For completeness all the relevant transformations are given below: $|s\rangle_{12+} \rightarrow |\phi\rangle_{3+}$

$$I \, |\phi\rangle_{3+} = (\alpha \, |0\rangle_3 + \beta \, |1\rangle_3) = |\phi\rangle_{3+} \qquad (23.24)$$

for $|s\rangle_{12-} \rightarrow |\phi\rangle_{3-}$

$$\sigma_z \, |\phi\rangle_{3-} = \sigma_z(\beta \, |1\rangle_3 - \alpha \, |0\rangle_3) = (\alpha \, |0\rangle_3 + \beta \, |1\rangle_3) = |\phi\rangle_{3+} \qquad (23.25)$$

for $|r\rangle_{12+} \rightarrow |\vartheta\rangle_{3+}$

$$\sigma_x \, |\vartheta\rangle_{3+} = \sigma_x(\alpha \, |1\rangle_3 + \beta \, |0\rangle_3) = (\alpha \, |0\rangle_3 + \beta \, |1\rangle_3) = |\phi\rangle_{3+} \qquad (23.26)$$

and for $|r\rangle_{12-} \rightarrow |\vartheta\rangle_{3-}$

$$i\sigma_y \, |\vartheta\rangle_{3-} = i\sigma_y(\beta \, |0\rangle_3 - \alpha \, |1\rangle_3) = (\alpha \, |0\rangle_3 + \beta \, |1\rangle_3) = |\phi\rangle_{3+}. \qquad (23.27)$$

Here, again it should be noted that the measured states by the emitter such as, $|s\rangle_{12+}$, $|s\rangle_{12-}$, $|r\rangle_{12+}$, and $|r\rangle_{12-}$, can be communicated to the receiver via a classical channel while the states $|\phi\rangle_{3+}$, $|\phi\rangle_{3-}$, $|\vartheta\rangle_{3+}$, and $|\vartheta\rangle_{3-}$ are transmitted quantum mechanically. The overall teleportation mechanics is captured in a simplified schematic presented in figure 23.1.

In summary, the transmitter starts with photon 1 in the state $|\phi\rangle_1 = (\alpha \, |0\rangle_1 + \beta \, |1\rangle_1)$ and proceeds to disassemble this state via a transformation that multiplies $|\phi\rangle_1$ with $|s\rangle_{23+}$.

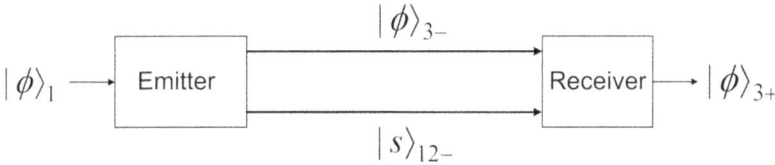

Figure 23.1. Simplified experimental schematics for quantum teleportation. The transmission channel for the $|\phi\rangle_{3-}$ state is a quantum channel while $|s\rangle_{12-}$ is sent to the receiver via a classical channel. The quantum states for $|\phi\rangle_1$ (the original state) and $|\phi\rangle_{3+}$ (the teleported state) are identical (see text).

Following this transformation, a series of additional states are created, such as $|\phi\rangle_{3+}$, $|\phi\rangle_{3-}$, $|\vartheta\rangle_{3+}$, and $|\vartheta\rangle_{3-}$. These states are transmitted quantum mechanically to the receiver. In turn, by using any of these states the receiver can recreate the original state now represented by photon 3 using the appropriate unitary transformation. In the example given here the unitary transformations are $I\,|\phi\rangle_{3+}$, $\sigma_z\,|\phi\rangle_{3-}$, $\sigma_x\,|\vartheta\rangle_{3+}$, and $i\sigma_y\,|\vartheta\rangle_{3-}$.

23.3 Technology

An interesting aspect of quantum teleportation is unitary transformations. These unitary transformations are performed using electro–optical modulators (Ma *et al* 2012). From a wider perspective, however, the quantum teleportation experiments integrate a variety of technologies including lasers, SPDC, half-wave plates, quarter-wave plates, polarizing beam splitters, fiber beam splitters, prisms, mirrors, telescopes, and single-photon detectors. For instance, the experiment performed by Ma *et al* (2012), over a distance of 143 km, included more than 60 optics and electro–optical components. Indeed, it is quite an achievement that despite the elaborate and complex nature of the experiments these researchers reported teleportation fidelities in the 0.76 − 0.80 range. Another quantum teleportation experiment, with a distance over 100 km, yielded a teleportation fidelity of just over 0.80 (Yin *et al* 2012).

Rapid advances in integrated optics, single-photon detection, and single-photon source technologies should improve the prospects and scope of the wonderful concept of quantum teleportation.

References

Bennett C H, Brassard G, Crépeau C, Jozsa R, Peres A and Wootters W K 1993 Teleporting an unknown quantum state via dual classical and Einstein–Podolsky–Rosen channels *Phys. Rev. Lett.* **70** 1895–9

Duarte F J 2014 *Quantum Optics for Engineers* (New York: CRC)

Feynman R P, Leighton R B and Sands M 1965 *The Feynman Lectures on Physics* vol III (Reading, MA: Addison-Wesley)

Fowles G R 1968 *Introduction to Modern Optics* (New York: Holt, Rinehart, and Winston)

Kim Y-H, Kulik S P and Shih Y 2001 Quantum teleportation of a polarization state with a complete Bell state measurement *Phys. Rev. Lett.* **86** 1370–3

Ma X *et al* 2012 Quantum teleportation over 143 kilometres using active feed-forward *Nature* **489** 269–73

Pirandola S, Eisert J, Weedbrook C, Furusawa A and Braunstein S L 2015 Advances in quantum teleportation *Nat. Photon.* **9** 641–52

Yin Y *et al* 2012 Quantum teleportation and entanglement distribution over 100-kilometre free-space channels *Nature* **488** 185–8

IOP Publishing

Fundamentals of Quantum Entanglement

F J Duarte

Chapter 24

Quantum entanglement and quantum computing

The application of the probability amplitude for quantum entanglement, $|\psi\rangle_- = (|x\rangle_1|y\rangle_2 - |y\rangle_1|x\rangle_2)$, to quantum computing is described. The Pauli gates and the Hadamard gate are expressed in terms of the four versions of the probability amplitude for quantum entanglement $|\psi\rangle_+$, $|\psi\rangle_-$, $|\psi\rangle^+$, and $|\psi\rangle^-$. Moreover, these probability amplitudes are expressed in terms of the identity matrix I and the Pauli matrices σ_x, σ_y, and σ_z.

24.1 Introduction

The first discussion of *quantum logic* appeared in the von Neumann book entitled *Mathematical Foundations of Quantum Mechanics*. In this book a quantum quantity is associated with the values of 0 and 1. The value of this quantity is 1 if it is verified, and 0 if it is not verified (von Neumann 1955).

In a more contemporaneous setting, quantum mechanics began its open literature association with computers, or computing, in the mid-1980s (Bennett 1982, Deutsch 1985, Feynman 1982, 1985, 1986, Peres 1985, Zurek 1984).

Among the topics that drove initial interest in this subject were the search for a Hamiltonian suitable for a universal computer, computational reversibility, and energy dissipation (Feynman 1985). The concept of quantum bit or *qbit* was introduced in the mid-1990s (Schumacher 1995).

Besides expected improvements in computational speed, reduced power consumption, and size minimization, quantum computing also promises computations based on quantum superposition and quantum entanglement, which are totally beyond the scope of classical universal computers.

Initially, however, quantum entanglement was not the focus of attention in the quantum computer research field. This concept was explicitly introduced around 1994 when equations of the form

doi:10.1088/2053-2563/ab2b33ch24

$$|\psi\rangle = \frac{1}{\sqrt{2}}(|x\rangle_1|y\rangle_2 \pm |y\rangle_1|x\rangle_2) \tag{24.1}$$

were highlighted and discussed along with concepts of quantum computing (Zeilinger *et al* 1994).

One further observation is that toward the year 2000, and beyond, in addition to the insertion of quantum entanglement, via equations of the form of the Pryce–Ward probability amplitude, the field of quantum computing began to reflect concepts originating in the philosophical path via the introduction of terminology like 'Bell states' and 'EPR pairs' (see, for instance, Nielsen and Chuang 2000).

The field of quantum computing, or quantum computers, is rather extensive, with more than 11 thousand articles published with the words 'quantum computing' in 2018 alone. Besides being the focus of a multitude of researchers, quantum computing is a rapidly evolving field with emerging technical advances aimed at achieving practical working quantum computers. The aim of this chapter is simply to provide a brief perspective on the presence of quantum entanglement principles in the area of quantum computation. For reviews of basic concepts in optical quantum computing the reader may consult Steane (1998) and Kok *et al* (2007).

24.2 Entropy

An important characteristic of quantum computers can be described from the perspective of the von Neumann entropy

$$S(\rho) = -\text{Tr}[\rho \ln \rho] \tag{24.2}$$

where ρ is the density matrix describing an ensemble of states of a relevant quantum system, and Tr is the trace of the matrix $[\rho \ln \rho]$ (von Neumann 1955). According to Schumacher (1995), 'the entropy $S(\rho)$ of a signal ensemble of pure states can be interpreted as the number of qbits per signal necessary to transpose it with near-perfect fidelity'. Transposition with high fidelity is central to quantum computing, thus making the reduction in computational noise crucial.

24.3 Qbits

Richard Feynman (1985, 1986) introduced the nexus between the concept of the *bit* and the quantum states $|1\rangle$ and $|0\rangle$. He then went on to propose that one bit can be represented by a single atom being either in the $|1\rangle$ state or the $|0\rangle$ state.

From the Dirac–Feynman principle (Dirac 1978, Feynman *et al* 1965), see chapters 2 and 17,

$$\langle\phi|\psi\rangle = \sum_{j=1}^{N}\langle\phi|j\rangle\langle j|\psi\rangle \tag{24.3}$$

for $j = 1, 2$

$$\langle\phi|\psi\rangle = \langle\phi|2\rangle\langle 2|\psi\rangle + \langle\phi|1\rangle\langle 1|\psi\rangle \tag{24.4}$$

and abstracting $\langle\phi|$

$$|\psi\rangle = |2\rangle\langle2|\psi\rangle + |1\rangle\langle1|\psi\rangle \qquad (24.5)$$

leads to

$$|\psi\rangle = C_2\,|2\rangle + C_1\,|1\rangle \qquad (24.6)$$

where

$$C_1 = \langle1|\psi\rangle \qquad (24.7)$$

$$C_2 = \langle2|\psi\rangle. \qquad (24.8)$$

Equation (24.6) is a general principle that means that a single state, or probability amplitude, can be expressed as a linear combination of two other states (Feynman *et al* 1965).

In the case of the $|1\rangle$ and $|0\rangle$ states, the combined probability amplitude of interest can be expressed as

$$|\psi\rangle = C_1\,|1\rangle + C_2\,|0\rangle \qquad (24.9)$$

where C_1 and C_2 are themselves probability amplitudes and therefore *complex numbers*. This is an example of a *superposition*. Once normalized, this probability amplitude can be expressed as

$$|\psi\rangle = \frac{1}{\sqrt{2}}(C_1\,|1\rangle + C_2\,|0\rangle). \qquad (24.10)$$

The $|1\rangle$ and $|0\rangle$ states are known in the literature as qbits, for 'quantum bits'; this terminology was introduced by Schumacher (1995).

The $|1\rangle$ and $|0\rangle$ states can also be entangled. In such a case, the probability amplitude can be expressed via equations like

$$|\psi\rangle = \frac{1}{\sqrt{2}}(|1\rangle|0\rangle - |0\rangle|1\rangle) \qquad (24.11)$$

or

$$|\psi\rangle = \frac{1}{\sqrt{2}}(|10\rangle - |01\rangle). \qquad (24.12)$$

Using the notation style of equation (24.11), the four possible probability amplitudes related to qbits $|1\rangle$ and $|0\rangle$ can be written as

$$|\psi\rangle = \frac{1}{\sqrt{2}}(|1\rangle|0\rangle \pm |0\rangle|1\rangle) \qquad (24.13)$$

$$|\psi\rangle = \frac{1}{\sqrt{2}}(|1\rangle|1\rangle \pm |0\rangle|0\rangle). \qquad (24.14)$$

These four states have become known as 'fully entangled Bell bases' (Zeilinger *et al* 1994).

For spin-1/2 particles, the individual states are also abbreviated as $|\uparrow\rangle$ and $|\downarrow\rangle$ so that the four corresponding probability amplitudes become

$$|\psi\rangle = \frac{1}{\sqrt{2}}(|\uparrow\rangle|\downarrow\rangle \pm |\downarrow\rangle|\uparrow\rangle) \tag{24.15}$$

$$|\psi\rangle = \frac{1}{\sqrt{2}}(|\uparrow\rangle|\uparrow\rangle \pm |\downarrow\rangle|\downarrow\rangle). \tag{24.16}$$

Alternatives of the probability amplitudes given in equations (24.15) and (24.16) are equations (24.13) and (24.14). These tend to be known in the literature as 'Bell states', named after Bell (1964), or 'EPR pairs', named after Einstein *et al* (1935) (see, for example, Nielsen and Chuang 2000).

24.4 Quantum entanglement and Pauli matrices

The probability amplitudes given in equations (24.13) and (24.14) can be expressed individually as

$$|\psi\rangle_+ = \frac{1}{\sqrt{2}}(|1\rangle|0\rangle + |0\rangle|1\rangle) \tag{24.17}$$

$$|\psi\rangle_- = \frac{1}{\sqrt{2}}(|1\rangle|0\rangle - |0\rangle|1\rangle) \tag{24.18}$$

$$|\psi\rangle^+ = \frac{1}{\sqrt{2}}(|1\rangle|1\rangle + |0\rangle|0\rangle) \tag{24.19}$$

$$|\psi\rangle^- = \frac{1}{\sqrt{2}}(|1\rangle|1\rangle - |0\rangle|0\rangle). \tag{24.20}$$

The quantum entanglement probability amplitudes can also be expressed in matrix form using the *direct vector product* (see appendix F) defined as (Ayres 1965)

$$|x\rangle|y\rangle = |x\rangle \cdot |y\rangle^T \tag{24.21}$$

which in explicit notation means

$$|x\rangle \cdot |y\rangle^T = \begin{pmatrix} x_1 \\ x_2 \end{pmatrix}\begin{pmatrix} y_1 \\ y_2 \end{pmatrix}^T = \begin{pmatrix} x_1y_1 & x_1y_2 \\ x_2y_1 & x_2y_2 \end{pmatrix}. \tag{24.22}$$

More specifically, using the direct vector product (see appendix F), equations (24.17)–(24.20) can be expressed in matrix form as (Duarte *et al* 2019)

$$|\psi\rangle^+ = 2^{-1/2}I \tag{24.23}$$

$$|\psi\rangle^- = 2^{-1/2}\sigma_z \tag{24.24}$$

$$|\psi\rangle_+ = 2^{-1/2}\sigma_x \tag{24.25}$$

$$|\psi\rangle_- = 2^{-1/2}i\sigma_y \tag{24.26}$$

where

$$I = \begin{pmatrix} 1 & 0 \\ 0 & 1 \end{pmatrix} \tag{24.27}$$

$$\sigma_x = \begin{pmatrix} 0 & 1 \\ 1 & 0 \end{pmatrix} \tag{24.28}$$

$$\sigma_y = \begin{pmatrix} 0 & -i \\ i & 0 \end{pmatrix} \tag{24.29}$$

$$\sigma_z = \begin{pmatrix} 1 & 0 \\ 0 & -1 \end{pmatrix} \tag{24.30}$$

I is the identity matrix while σ_x, σ_y, and σ_z are the corresponding Pauli matrices.

In general, this means that any optical component associated with a 2×2 matrix operator operating on one of the entangled vector states can yield a new transformed state. For instance, if an optical component is represented by the matrix operator B, then where $|\xi\rangle$ is a new or transformed state

$$B|\psi\rangle_+ \rightarrow |\xi\rangle \tag{24.31}$$

If the entangled photons are made to be incident on a generalized polarization rotator (see appendix E, or Duarte 2014) represented by the matrix

$$R = \begin{pmatrix} 0 & 1 \\ 1 & 0 \end{pmatrix} \tag{24.32}$$

then it can be shown that (Duarte *et al* 2019)

$$R|\psi\rangle^+ = |\psi\rangle_+ = 2^{-1/2}\sigma_x \tag{24.33}$$

$$R|\psi\rangle^- = -|\psi\rangle_- = -2^{-1/2}i\sigma_y \tag{24.34}$$

$$R\,|\psi\rangle_+ = |\psi\rangle^+ = 2^{-1/2}I \tag{24.35}$$

$$R|\psi\rangle_- = -|\psi\rangle^- = -2^{-1/2}\sigma_z. \tag{24.36}$$

This thus demonstrates an inherent capability for mathematical operations. This simple concept has significant practical implications. It means that beams of entangled photons interacting with optical apparatus configured with optical elements capable of transforming the entangled states can be used to perform optical quantum computations.

In addition to the polarization rotator, other optical components of interest in optical computers are the beam splitter and the Mach–Zehnder interferometer. These are considered in appendix D.

24.5 Pauli matrices and quantum entanglement

It can also be shown, using the direct vector product technique, that (Duarte *et al* 2019)

$$\sigma_x = 2^{1/2}|\psi\rangle_+ \tag{24.37}$$

$$i\sigma_y = 2^{1/2}|\psi\rangle_- \tag{24.38}$$

$$I = 2^{1/2}|\psi\rangle^+ \tag{24.39}$$

$$\sigma_z = 2^{1/2}|\psi\rangle^-. \tag{24.40}$$

These identities show yet an additional avenue to derive the quantum entanglement probability amplitudes provided that the Pauli matrices are arrived at independently and from first principles.

24.6 Quantum gates

In this section the characteristics of some basic quantum gates linked to the probability amplitude of quantum entanglement are introduced.

In the previous section the reader should have noticed that

$$R = \sigma_x = \begin{pmatrix} 0 & 1 \\ 1 & 0 \end{pmatrix}. \tag{24.41}$$

In other words, a polarization rotator behaves mathematically exactly as the σ_x Pauli matrix. Indeed, the Pauli σ_x matrix is the mathematical realization of a quantum NOT gate whose truth table is given below:

$$
\begin{array}{cc}
a & a' \\
0 & 1 \\
1 & 0.
\end{array}
$$

Using the Feynman approach, the quantum NOT gate is illustrated pictorially in figure 24.1. Furthermore, the observation summarized in equation (24.41) implies that any optical element that can be related to one of Pauli's matrices is capable of implementing a gate for a quantum computer.

It should be noticed that in practice an optical element that induces a $\pi/2$ polarization can be realized using a half-wave plate, a wavelength-specific Fresnel rhomb, or a broadband collinear prismatic rotator (Duarte 1989, 2014). A quantum NOT gate was demonstrated by Pelliccia *et al* (2003) via polarization methods.

In classical universal computers, logical gates such as AND, NAND, OR, NOR, and NOT form the bases for logical operations. In quantum computing, a gate that

(a)

(b)

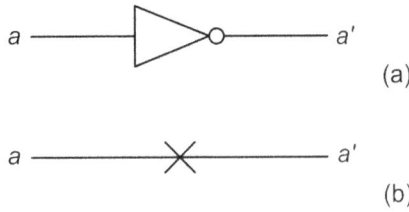

Figure 24.1. Generic versions of the NOT gate: (a) classical and (b) quantum.

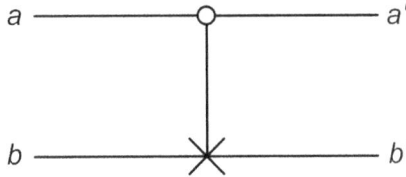

Figure 24.2. Generic version of the quantum CNOT gate.

was of particular interest to Feynman (1985) was the controlled NOT gate or CNOT gate. The truth table for the CNOT gate is given by (Feynman 1985)

$$
\begin{array}{cccc}
a & b & a' & b' \\
0 & 0 & 0 & 0 \\
0 & 1 & 0 & 1 \\
1 & 0 & 1 & 1 \\
1 & 1 & 1 & 0.
\end{array}
$$

In the truth table of the CNOT gate, the value of b' is changed *if and only if* the value of $a = 1$. The pictorial representation of the CNOT gate is illustrated in figure 24.2. The interest in the CNOT gate arrives from the fact that this is a fundamental gate for the generation of quantum entanglement states.

The first realization of a quantum CNOT gate was reported by Wineland and colleagues (Monroe *et al* 1995).

24.6.1 Pauli gates

There are three Pauli gates (Moore and Nilsson 2001, Childs 2005) and they are identified as depicted in figure 24.3. Each Pauli matrix (σ_x, σ_y, σ_z) provides the mathematical representation for each of the corresponding gates. Each of the Pauli gates act on a single qbit. The Pauli-X gate is equivalent to the NOT gate, and physically, as already indicated in equation (24.41), it induces a polarization rotation of $\pi/2$ radians. In other words, it converts $|x\rangle$ to $|y\rangle$, or $|y\rangle$ to $|x\rangle$. Specifically,

$$
\begin{pmatrix} 0 & 1 \\ 1 & 0 \end{pmatrix} \begin{pmatrix} 1 \\ 0 \end{pmatrix} = \begin{pmatrix} 0 \\ 1 \end{pmatrix}
\tag{24.42}
$$

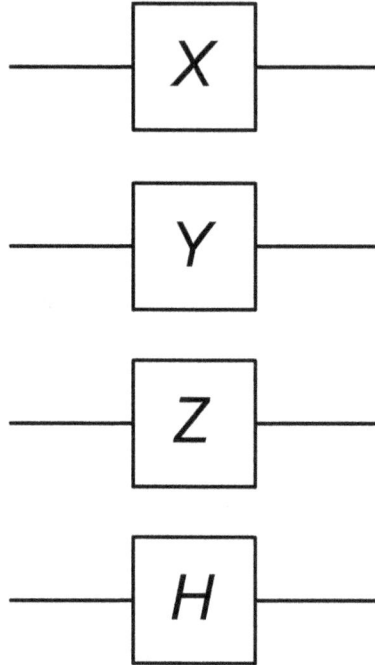

Figure 24.3. Generic symbol representation for the Pauli gates corresponding to σ_x, σ_y, and σ_z. The Hadamard H gate is also included.

$$\begin{pmatrix} 0 & 1 \\ 1 & 0 \end{pmatrix}\begin{pmatrix} 0 \\ 1 \end{pmatrix} = \begin{pmatrix} 1 \\ 0 \end{pmatrix} \tag{24.43}$$

which can be summarized as

$$\sigma_x |1\rangle \rightarrow |0\rangle \tag{24.44}$$

$$\sigma_x |0\rangle \rightarrow |1\rangle. \tag{24.45}$$

The Pauli-Y gate performs the following transformations

$$\begin{pmatrix} 0 & -i \\ i & 0 \end{pmatrix}\begin{pmatrix} 1 \\ 0 \end{pmatrix} = \begin{pmatrix} 0 \\ i \end{pmatrix} \tag{24.46}$$

$$\begin{pmatrix} 0 & -i \\ i & 0 \end{pmatrix}\begin{pmatrix} 0 \\ 1 \end{pmatrix} = \begin{pmatrix} -i \\ 0 \end{pmatrix} \tag{24.47}$$

that can be summarized as

$$\sigma_y |1\rangle \rightarrow i|0\rangle \tag{24.48}$$

$$\sigma_y |0\rangle \rightarrow -i|1\rangle. \tag{24.49}$$

The Pauli-Z gate conducts the transformations

$$\begin{pmatrix} 1 & 0 \\ 0 & -1 \end{pmatrix}\begin{pmatrix} 1 \\ 0 \end{pmatrix} = \begin{pmatrix} 1 \\ 0 \end{pmatrix} \tag{24.50}$$

$$\begin{pmatrix} 1 & 0 \\ 0 & -1 \end{pmatrix}\begin{pmatrix} 0 \\ 1 \end{pmatrix} = \begin{pmatrix} 0 \\ -1 \end{pmatrix} \tag{24.51}$$

which can be summarized as

$$\sigma_z |1\rangle \rightarrow |1\rangle \tag{24.52}$$

$$\sigma_z |0\rangle \rightarrow -|0\rangle. \tag{24.53}$$

24.6.2 The Hadamard gate

The Hadamard gate is related to the matrices of the same name, which are designated after the mathematician J Hadamard. The matrix of interest is

$$H = 2^{-1/2}\begin{pmatrix} 1 & 1 \\ 1 & -1 \end{pmatrix} \tag{24.54}$$

which can be derived from the following operation involving Pauli matrices:

$$H = 2^{-1/2}(\sigma_x + \sigma_z). \tag{24.55}$$

This gate performs the following transformations:

$$2^{-1/2}\begin{pmatrix} 1 & 1 \\ 1 & -1 \end{pmatrix}\begin{pmatrix} 1 \\ 0 \end{pmatrix} = 2^{-1/2}\begin{pmatrix} 1 \\ 1 \end{pmatrix} = 2^{-1/2}\left(\begin{pmatrix} 1 \\ 0 \end{pmatrix} + \begin{pmatrix} 0 \\ 1 \end{pmatrix}\right) \tag{24.56}$$

$$2^{-1/2}\begin{pmatrix} 1 & 1 \\ 1 & -1 \end{pmatrix}\begin{pmatrix} 0 \\ 1 \end{pmatrix} = 2^{-1/2}\begin{pmatrix} 1 \\ -1 \end{pmatrix} = 2^{-1/2}\left(\begin{pmatrix} 1 \\ 0 \end{pmatrix} - \begin{pmatrix} 0 \\ 1 \end{pmatrix}\right) \tag{24.57}$$

which can be summarized as

$$H|1\rangle \rightarrow 2^{-1/2}(|1\rangle + |0\rangle) \tag{24.58}$$

$$H|0\rangle \rightarrow 2^{-1/2}(|1\rangle - |0\rangle). \tag{24.59}$$

24.7 The Hadamard matrix and quantum entanglement

It can also be shown that the Hadamard matrix operation on the probability amplitudes for quantum entanglement yield the following mathematical identities (Duarte and Taylor 2019):

$$H|\psi\rangle_+ = 2^{-1/2}(|\psi\rangle^+ + |\psi\rangle_-) \tag{24.60}$$

$$H|\psi\rangle_- = 2^{-1/2}(|\psi\rangle_+ - |\psi\rangle^-) \tag{24.61}$$

$$H|\psi\rangle^+ = 2^{-1/2}(|\psi\rangle_+ + |\psi\rangle^-) \tag{24.62}$$

$$H|\psi\rangle^- = 2^{-1/2}(|\psi\rangle^+ - |\psi\rangle_-) \tag{24.63}$$

which are equivalent to (Duarte and Taylor 2019)

$$H|\psi\rangle_+ = 2^{-1}(I + i\sigma_y) \tag{24.64}$$

$$H|\psi\rangle_- = 2^{-1}(\sigma_x - \sigma_z) \tag{24.65}$$

$$H|\psi\rangle^+ = 2^{-1}(\sigma_x + \sigma_z) \tag{24.66}$$

$$H|\psi\rangle^- = 2^{-1}(I - i\sigma_y). \tag{24.67}$$

Finally, from equation (24.55) and equations (24.29) and (24.31) (see appendix F),

$$H = (|\psi\rangle_+ + |\psi\rangle^-) \tag{24.68}$$

which can also be derived directly from equations (24.17) and (24.20).

24.8 Multiple entangled states

Binary computing is based on the power of 2, or $n = N = 2^1, 2^2, 2^3 \ldots 2^r$. For the case $n = N = 2^2$, the states $|0\rangle$ and $|1\rangle$ can form a series of combined states such as $|1\rangle|1\rangle|1\rangle|1\rangle, |1\rangle|1\rangle|0\rangle|1\rangle, |1\rangle|1\rangle|1\rangle|0\rangle, |1\rangle|1\rangle|0\rangle|0\rangle, |1\rangle|0\rangle|1\rangle|1\rangle, |1\rangle|0\rangle|0\rangle|1\rangle, |1\rangle|0\rangle|1\rangle|0\rangle, |1\rangle|0\rangle|0\rangle|0\rangle, |0\rangle|1\rangle|1\rangle|1\rangle, |0\rangle|1\rangle|1\rangle|0\rangle, |0\rangle|1\rangle|0\rangle|1\rangle, |0\rangle|1\rangle|0\rangle|0\rangle, |0\rangle|0\rangle|1\rangle|1\rangle, |0\rangle|0\rangle|1\rangle|0\rangle, |0\rangle|0\rangle|0\rangle|1\rangle$, and $|0\rangle|0\rangle|0\rangle|0\rangle$. However, only four of these combined states adhere to the requirement of inter-pair orthogonality relevant to quantum entanglement:

$$|C\rangle_1 = |1\rangle|0\rangle|0\rangle|1\rangle \tag{24.69}$$

$$|C\rangle_2 = |1\rangle|0\rangle|1\rangle|0\rangle \tag{24.70}$$

$$|C\rangle_3 = |0\rangle|1\rangle|0\rangle|1\rangle \tag{24.71}$$

$$|C\rangle_4 = |0\rangle|1\rangle|1\rangle|0\rangle. \tag{24.72}$$

Thus, the corresponding combined and normalized probability amplitudes become (Duarte 2015, 2016)

$$|\psi\rangle_I = \frac{1}{\sqrt{4}}(|C\rangle_1 + |C\rangle_2 + |C\rangle_3 + |C\rangle_4) \tag{24.73}$$

$$|\psi\rangle_{II} = \frac{1}{\sqrt{4}}(|C\rangle_1 + |C\rangle_2 - |C\rangle_3 - |C\rangle_4) \tag{24.74}$$

$$|\psi\rangle_{III} = \frac{1}{\sqrt{4}}(|C\rangle_1 - |C\rangle_2 + |C\rangle_3 - |C\rangle_4) \tag{24.75}$$

$$|\psi\rangle_{IV} = \frac{1}{\sqrt{4}}(|C\rangle_1 - |C\rangle_2 - |C\rangle_3 + |C\rangle_4). \tag{24.76}$$

The principles of $n = N = 2^3$, 2^4 quantum entanglement, which are applicable to binary quantum computing, are described in chapter 18 and also discussed by Duarte and Taylor (2017). The principles applicable to $n = N = 3$ and $n = N = 6$ were outlined in chapter 19.

24.9 Technology

Two important areas of technological development in optical quantum computing are high-fidelity computations, based on either quantum superposition or quantum entanglement, and the capability to handle large numbers of qbits. The aim here is to generate large numbers of qbits, 10^3, 10^6, or higher, in an extremely low-noise environment so that single-photon registrations can be identified with high fidelity. This field is rapidly evolving while incorporating new single-photon-pair sources and improved single-photon detection technology. Once a large number of qbits, for instance in the $10^3 \leqslant N_{qbit} \leqslant 10^6$ range, is generated routinely with high fidelity, the task will be to achieve a miniaturization that will translate the technology from a table-top environment to an on-chip realization.

At present, this field is rapidly advancing with numerous competing approaches. This exciting and fast-moving state of development renders any attempt to summarize the technology obsolete almost as soon as it is written.

References

Ayres F 1965 *Modern Algebra* (New York: McGraw-Hill)

Bell J S 1964 On the Einstein–Podolsky–Rosen paradox *Physics* **1** 195–200

Bennett C H 1982 Thermodynamics of computation—a review *Int. J. Theor. Phys.* **21** 905–40

Childs A M 2005 Secure assisted quantum computation *Quantum Inf. Comput.* **5** 456

Dirac P A M 1978 *The Principles of Quantum Mechanics* 4th edn (Oxford: Oxford University Press)

Deutsch D 1985 Quantum theory, the Church–Turing principle and the universal quantum computer *Proc. R. Soc. London A Math. Phys. Sci.* **400** 97–117

Duarte F J 1989 Optical device for rotating the polarization of a light beam *US Patent 4822150*

Duarte F J 2014 *Quantum Optics for Engineers* (New York: CRC)

Duarte F J 2015 *Tunable Laser Optics* 2nd edn (New York: CRC)

Duarte F J 2016 Secure space-to-space interferometric communications and its nexus to the physics of quantum entanglement *Appl. Phys. Rev.* **3** 041301

Duarte F J and Taylor T S 2017 Quantum entanglement probability amplitudes in multiple channels: an interferometric approach *Optik* **139** 222–30

Duarte F J and Taylor T S 2019 The Hadamard matrix and the probability amplitude for quantum entanglement *Unpublished*

Duarte F J, Taylor T S and Slaten J C 2019 On the probability amplitude of quantum entanglement and the Pauli matrices *Unpublished*

Einstein A, Podolsky B and Rosen N 1935 Can quantum mechanical description of physical reality be considered complete? *Phys. Rev.* **47** 777–80

Feynman R P 1982 Simulating physics with computers *Int. J. Theor. Phys.* **21** 467–88

Feynman R P 1985 Quantum mechanical computers *Opt. News* **11** 11–20

Feynman R P 1986 Quantum mechanical computers *Found. Phys.* **16** 507–31

Feynman R P, Leighton R B and Sands M 1965 *The Feynman Lectures on Physics* vol III (Reading, MA: Addison-Wesley)

Kok P, Munro W J, Nemoto K, Ralph T C, Dowling J P and Milburn G J 2007 Linear optical quantum computing with photonic qbits *Rev. Mod. Phys.* **79** 135–74

Monroe C, Meekhof D M, King B E, Itano W M and Wineland D J 1995 Demonstration of a fundamental quantum logic gate *Phys. Rev. Lett.* **75** 4714–7

Moore C and Nilsson M 2001 Parallel quantum computation and quantum codes *SIAM J. Comput.* **31** 799–815

Nielsen M A and Chuang I L 2000 *Quantum Computation and Quantum Information* (Cambridge: Cambridge University Press)

Pelliccia D, Schettini V, Sciarrino F, Sias C and De Martini F 2003 Contextual realization of the universal quantum cloning machine and of the universal-NOT gate by quantum-injected optical parametric amplification *Phys. Rev.* A **68** 042306

Peres A 1985 Reversible logic and quantum computers *Phys. Rev.* A **32** 3266–76

Schumacher B 1995 Quantum coding *Phys. Rev. Lett.* **51** 2738–47

Steane A 1998 Quantum computing *Rep. Prog. Phys.* **61** 117–74

von Neumann J 1955 *Mathematical Foundations of Quantum Mechanics* (Princeton, NJ: Princeton University Press) Note: this is a translation of the original title published in German in 1932

Zeilinger A, Bernstein H J and Horne M A 1994 Information transfer with two-state two-particle quantum systems *J. Mod. Opt.* **41** 2375–84

Zurek W H 1984 Reversibility and stability of information processing systems *Phys. Rev. Lett.* **53** 391–4

Chapter 25

Space-to-space and space-to-Earth communications via quantum entanglement

Space-to-space and space-to-Earth communications in the context of the probability amplitude for quantum entanglement are discussed.

25.1 Introduction

The physics of quantum entanglement led to quantum cryptography, and quantum cryptography led to the concept of long-distance secure communications, and eventually to long-distance free-space secure communications and space-to-space cryptographic communications. This is a subject still in its early infancy.

Historically, one of the first explicit links between quantum entanglement and space-to-space communications was made by Aspelmeyer *et al* (2003). More recently, this subject has been addressed, from an alternative perspective, by Duarte and Taylor (2015) and Duarte (2016). For an introduction to and review of concepts on space-to-space laser communications, the reader may refer to Lambert and Casey (1995).

In this chapter, a brief introduction to salient features of space-to-space and space-to-Earth quantum entanglement communications is given.

25.2 Space-to-space configurations

The beauty of space-to-space laser, or optical, communications is that physicists and engineers do not have to deal with atmospheric turbulence. In this regard, the vacuum of space is an ideal propagation medium. However, due to the large distances involved, the technologies of acquisition, tracking, and pointing assume an augmented importance. For reviews on these technologies, the reader is referred to Lambert and Casey (1995) and Kaushal and Kaddoum (2017).

In this section attention is limited to simple space-to-space configurational arrangements inherently relevant to the optical architectures of quantum entanglement systems. The main concern in these long-distance arrangements, which will not be quantified here but are considered by Aspelmeyer *et al* (2003), are losses that increase as the aperture diameter of the transmitter and receiver optics decreases.

First, communications between two parties, A and B, are considered. Traditionally in communications systems the transmitting party A, creates, generates, and transmits the message, and thus it is close to the transmission source. The receiving party B can be either a passive receiver, as outlined in figure 25.1, or could also have transmitting capabilities, as outlined in figure 25.2. In cryptographic quantum entanglement arrangements, B needs some sort of transmitting capability so that the quantum code, or key, can be verified. Furthermore, in quantum entanglement configurations, the source S can be remote to both A and B, as illustrated in figure 25.3. This is a feature that became apparent from the very first schematics on quantum entanglement (Pryce and Ward 1947, Ward 1949).

A likely scenario that will have to be met by quantum entanglement cryptography satellites is that of a series of satellites intercommunicating with each other either in low Earth orbit (LEO) or geosynchronous Earth orbit (GEO), as outlined in figure 25.4. Depending on the orbit, intersatellite array separation distances D of interest are in the $1 \times 10^6 \leqslant D \leqslant 10 \times 10^6$ m range (Duarte 2016). Quantum entanglement cryptography systems will have to adapt to satisfy satellite network array configurations and practical intersatellite distances.

25.3 The space-to-Earth experiment

A landmark space-to-Earth quantum entanglement experiment was reported by Yin *et al* (2017). This experimental configuration followed the pattern outlined in figure 25.3 but with both A and B located on the ground. The terrestrial distance between A and B was reported to be 1203 km, the distance between S and A was reported to vary in the $560 \leqslant D_{SA} \leqslant 1700$ km range, and the distance between S and B was reported to vary in the $545 \leqslant D_{SB} \leqslant 1680$ km range.

Figure 25.1. Diagram depicting laser, or photon, communications between a transmitting satellite A and a receiving satellite B.

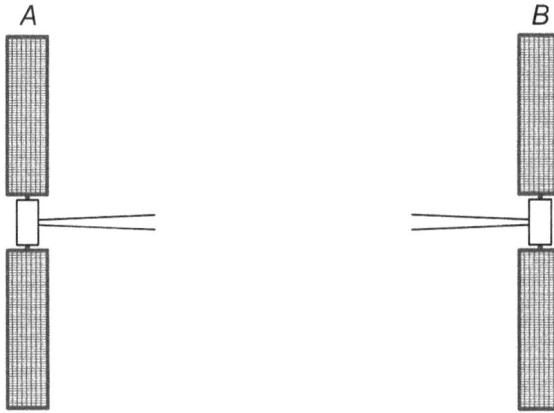

Figure 25.2. Diagram depicting laser, or photon, communications between satellite *A* and a receiving satellite *B*. Both satellites are capable of transmission and reception.

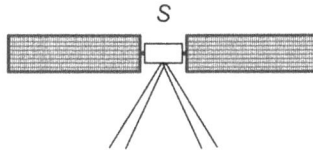

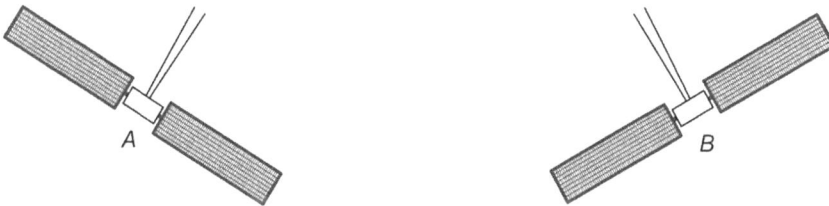

Figure 25.3. Diagram depicting entangled photon communications between satellite *A* and satellite *B*. The source of the entangled photon pairs is at satellite *S*.

The source at the satellite is a 30 mW diode laser emitting at $\lambda \approx 405$ nm, with a linewidth of $\Delta \nu = 160$ MHz, utilized to excite a $KTiOPO_4$ (PPKTP) crystal within a Sagnac interferometer configuration capable of generating 5.9×10^6 photon pairs per second with a reported fidelity of 0.907 ± 007 at $\lambda \approx 810$ nm (Yin *et al* 2017). The two entangled beams are emitted via 300 mm and 180 mm aperture Cassegrain

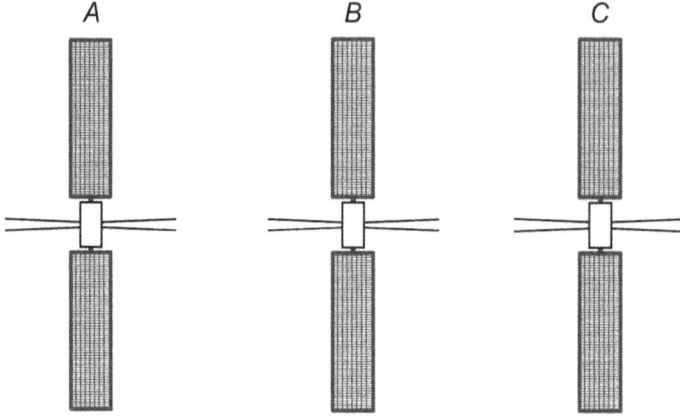

Figure 25.4. Simplified diagram depicting a satellite array at either LEO or GEO. Satellites have both transmitting and receiving capabilities.

telescopes. The receiving telescopes were also of the Cassegrain class, with apertures of 1200 mm and 1800 mm (Yin *et al* 2017). Overall, this experiment comprises more than 70 optical and electro–optical components.

The entangled polarization state generated by the emitter was of the form

$$|\psi\rangle = \frac{1}{\sqrt{2}}(|H\rangle_1 |V\rangle_2 + |V\rangle_1 |H\rangle_2) \qquad (25.1)$$

and the integrity of the transmission was corroborated using Bell's inequality (Bell 1964), as expressed in chapter 10

$$\Sigma_P = |P(\varphi_1, \varphi_2) - P(\varphi_1, \varphi_2')| + |P(\varphi_1', \varphi_2') + P(\varphi_1', \varphi_2)| \leqslant 2. \qquad (25.2)$$

The angles φ_1 and φ_2 were selected randomly from the following sets: $(0, \pi/8)$, $(0, 3\pi/8)$, $(\pi/4, \pi/8)$, and $(\pi/4, 3\pi/8)$. Following 1167 trials it was determined that $\Sigma_P = 2.37 \pm 0.09$ (Yin *et al* 2017), thus violating Bell's inequality and pointing toward no third-party intrusion in the transmission.

It is interesting to note the apparent minimal adverse effects from atmospheric turbulence and absorption on these reported measurements. However, this is not entirely unexpected as minimal turbulence effects on the transmission of interferometric signals have been previously reported from measurements in *terrestrial* environments noted for their average high humidity and turbulence (Duarte *et al* 2011).

25.4 Further horizons

The physics of quantum entanglement has been known since 1947 (Pryce and Ward 1947) and experimental confirmations have been forthcoming since 1948 (Hanna 1948, Bleuler and Bradt 1948, Wu and Shaknov 1950). In this regard, the field has moved beyond physics and into the realm of engineering.

From a technological perspective, space-to-space, space-to-Earth, and Earth-to-space quantum cryptography communications require advances in sources of entangled photon pairs and in electro–optical integration in order to minimize the number of components necessary to achieve secure communications in an efficient and reliable fashion. Conceptual advances in the physics of quantum entanglement may also lead to the simplification of and configurational improvements in optical systems.

References

Aspelmeyer M, Jennewein T, Pfennigbauer M, Leeb W R and Zeilinger A 2003 Long-distance quantum communication with entangled photons using satellites *IEEE J. Select. Top. Quantum Electron.* **9** 1541–51

Bell J S 1964 On the Einstein–Podolsky–Rosen paradox *Physics* **1** 195–200

Bleuler E and Bradt H L 1948 Correlation between the states of polarization of the two quanta of annihilation radiation *Phys. Rev.* **73** 1398

Duarte F J 2016 Secure space-to-space interferometric communications and its nexus to the physics of quantum entanglement *Appl. Phys. Rev.* **3** 041301

Duarte F J, Taylor T S, Black A M, Davenport W E and Varmette V G 2011 N-slit interferometer for secure free-space optical communications: 527 m intra interferometric path length *J. Opt.* **13** 035710

Duarte F J and Taylor T S 2015 Quantum entanglement physics secures space-to-space interferometric communications *Laser Focus World* **51** 54–8

Hanna R C 1948 Polarization of annihilation radiation *Nature* **162** 332

Kaushal H and Kaddoum G 2017 Optical communications in space: challenges and mitigation techniques *IEEE Commun. Surv. Tutor.* **19** 57–96

Lambert S G and Casey W L 1995 *Laser Communications in Space* (Boston, MA: Artech House)

Pryce M L H and Ward J C 1947 Angular correlation effects with annihilation radiation *Nature* **160** 435

Ward J C 1949 *Some Properties of the Elementary Particles* (Oxford: Oxford University Press)

Wu C S and Shaknov I 1950 The angular correlation of scattered annihilation radiation *Phys. Rev.* **77** 136

Yin J *et al* 2017 Satellite-based entanglement distribution over 1200 kilometers *Science* **356** 1140–4

IOP Publishing

Fundamentals of Quantum Entanglement

F J Duarte

Chapter 26

Space-to-space quantum interferometric communications: an alternative to quantum entanglement communications?

Space-to-space communications based on the generation of N-slit quantum interferometric characters (QICs), originating from the Dirac–Feynman probability amplitude principle, are described. It is emphasized that the security integrity of this approach is inherently protected by the probability amplitude principle and does not require the generation of a key or a code, nor does it need to use Bell's theorem.

26.1 Introduction

Previously, in chapter 17, it was shown how the probability amplitude for two quanta propagating in opposite directions, with entanglement polarizations (Pryce and Ward 1947, Ward 1949)

$$|\psi\rangle = \frac{1}{\sqrt{2}}(|x\rangle_1 |y\rangle_2 - |y\rangle_1 |x\rangle_2)$$

(26.1)

can be derived from the very foundations of quantum mechanics utilizing the Dirac–Feynman principle (Dirac 1939, 1978, Feynman *et al* 1965)

$$\langle x|s\rangle = \sum_{j=1}^{N}\langle x|j\rangle\langle j|s\rangle.$$

(26.2)

In this chapter, it is shown how the generalized N-slit probability equations derived from this principle can be used to generate QICs for long-distance communications and how quantum mechanics inherently protects these interferometric signals from being harvested by a third-party intruder. In contrast to quantum entanglement methodology, this approach *does not* require a key or code, and the meaning of the QICs can be made public if the communicating parties, A or B, so desire. There is also no need to use Bell's theorem (Bell 1964) to verify security. An additional advantage over quantum entanglement communications is that although communications can use single-photon sources, the equations equally apply to populations of indistinguishable photons as available from single-transverse-mode single-longitudinal-mode narrow-linewidth lasers. This enormously improves the efficiency of the communications and the signal-to noise ratio, thus relaxing the requirements of detection technology. One disadvantage of the interferometric approach is related to beam divergence characteristics that might limit its applicability over very long space-to-space distances. Furthermore, this approach is only applicable to free-space communications and is not suitable to transmissions via optical fibers.

This approach to free-space communications was first mentioned in the literature by Duarte (2002), and further details have been published in Duarte (2005, 2014, 2016), Duarte *et al* (2010), Duarte *et al* (2011) and Duarte and Taylor (2015).

26.2 The generalized N-slit quantum interference equations

As already explained in chapter 2, the Dirac–Feynman probability amplitude

$$\langle x|s \rangle = \sum_{j=1}^{N} \langle x|j \rangle \langle j|s \rangle$$

leads directly, via Born's rule (Born 1926), to the generalized probability in one dimension:

$$\langle x|s \rangle \langle x|s \rangle^* = \left(\sum_{j=1}^{N} \langle x|j \rangle \langle j|s \rangle \right) \left(\sum_{j=1}^{N} \langle x|j \rangle \langle j|s \rangle \right)^*. \tag{26.3}$$

These equations apply to the propagation of a single photon, or a population of indistinguishable photons, from a source s, via an array of N slits j, to an interference surface x, as illustrated in figure 26.1. Representing the probability amplitudes with *wave functions of ordinary wave optics* (Dirac 1978) such as

$$\langle j|s \rangle = \Psi(r_{j,s})e^{-i\theta_j} \tag{26.4}$$

$$\langle x|j \rangle = \Psi(r_{x,j})e^{-i\phi_j} \tag{26.5}$$

leads to the generalized interferometric equation in complex form (Duarte and Paine 1989, Duarte 1991, 1993):

$$|\langle x|s \rangle|^2 = \sum_{j=1}^{N} \Psi(r_j) \sum_{m=1}^{N} \Psi(r_m)e^{i(\Omega_m - \Omega_j)} \tag{26.6}$$

which can also be expressed as

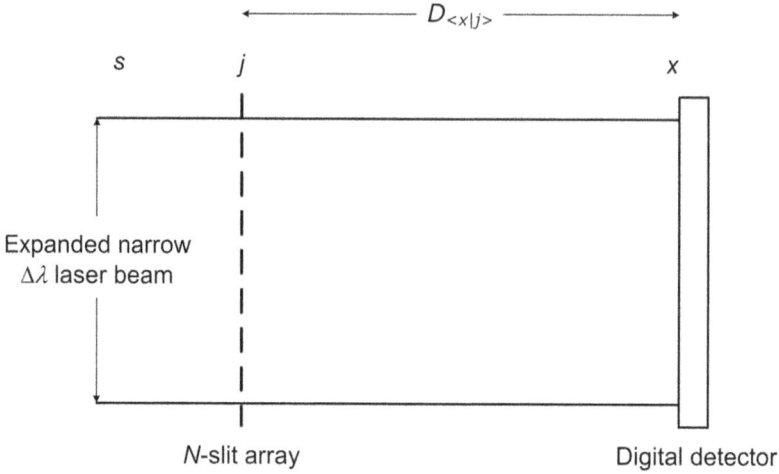

Figure 26.1. Generic diagram of an N-slit laser interferometer.

$$|\langle x|s\rangle|^2 = \sum_{j=1}^{N}\Psi(r_j)^2 + 2\sum_{j=1}^{N}\Psi(r_j)\left(\sum_{m=j+1}^{N}\Psi(r_m)\cos(\Omega_m - \Omega_j)\right). \qquad (26.7)$$

The generalized interferometric equation in its complex form, as given in equation (26.6), can be expressed in two dimensions as (Duarte 1995, 2014)

$$|\langle x|s\rangle|^2 = \sum_{z=1}^{N}\sum_{y=1}^{N}\Psi(r_{zy})\sum_{q=1}^{N}\sum_{p=1}^{N}\Psi(r_{pq})e^{i(\Omega_{qp}-\Omega_{zy})} \qquad (26.8)$$

and in three dimensions as (Duarte 1995, 2014)

$$|\langle x|s\rangle|^2 = \sum_{z=1}^{N}\sum_{y=1}^{N}\sum_{x=1}^{N}\Psi(r_{zyx})\sum_{q=1}^{N}\sum_{p=1}^{N}\sum_{r=1}^{N}\Psi(r_{qpr})e^{i(\Omega_{qpr}-\Omega_{zyx})}. \qquad (26.9)$$

26.3 The generation and transmission of interferometric characters

The N-slit interferometer is comprised of a single-transverse-mode narrow-linewidth laser capable of generating populations of indistinguishable photons. The laser is then followed by a one-dimensional non-focusing beam expander such as a multiple-prism beam expander (Duarte and Piper 1982, Duarte 2015). This combined laser beam–expander system becomes the source of indistinguishable photons s. The expanded narrow-linewidth laser beam then illuminates an array of N slits (j). The transmitter A controls the number of slits being illuminated and thus the interferometric character to be transmitted. The receiver B is at the interferometric plane x. The intra–interferometric distance between j and the interference plane x is denoted as $D_{\langle x|j\rangle}$, and the whole system is depicted in figure 26.2.

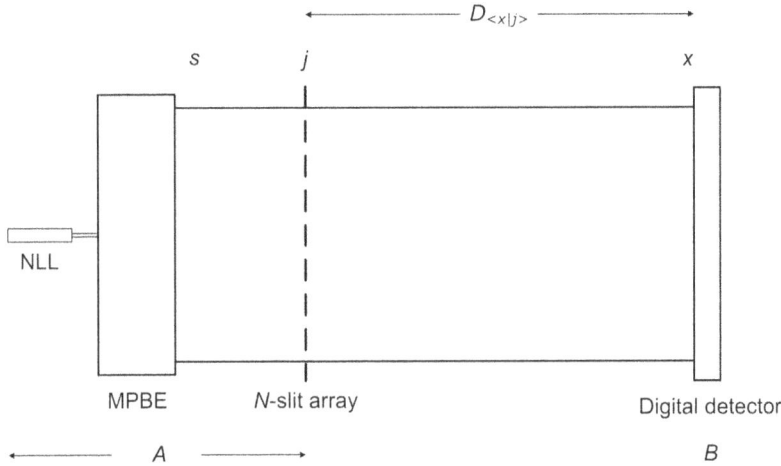

Figure 26.2. The N-slit laser interferometer depicting the transmitter stage A and the receiver stage B. NLL stands for narrow-linewidth laser and **MPBE** stands for multiple-prism beam expander.

The interferometric characters are generated by assigning a number of slits to a given alpha-numerical character. For instance, $N = 2$ is a, $N = 3$ is b, $N = 4$ is c ... $N = 26$ is z (Duarte 2002, 2015). Any attempt by a third party to obtain information from the propagating interferometric character has a catastrophic effect and destroys the information being sent from A to B and also renders useless the information received by the intruder or eavesdropper (Duarte 2002). There is no need for a quantum key or code.

If spying is detected, the only action needed is for B to communicate to A that an interception has occurred. The simplicity of the system also allows B to have interferometric transmitting capabilities. A protocol to detect possible interception, and re-transmission, of the QICs by cloaked intruders is discussed by Duarte (2016).

26.4 The inherent quantum security mechanism

Probability amplitudes are inherently extremely sensitive to even minute physical transformations. If a further intermediate plane, such as a beam splitter k, is introduced between j and x as illustrated in figure 26.3, then the probability amplitude undergoes a severe alteration and becomes

$$\langle x|s \rangle = \sum_{k=1}^{N}\sum_{j=1}^{N}\langle x|k\rangle\langle k|j\rangle\langle j|s\rangle. \tag{26.10}$$

This probability amplitude introduces two propagation distances $D_{\langle x|k\rangle}$ and $D_{\langle k|j\rangle}$ rather than just $D_{\langle x|j\rangle}$. Even if the beam splitter is replaced by a single strand of a transparent microscopic fiber, such as a spider web silk fiber, the probability amplitude is transformed to

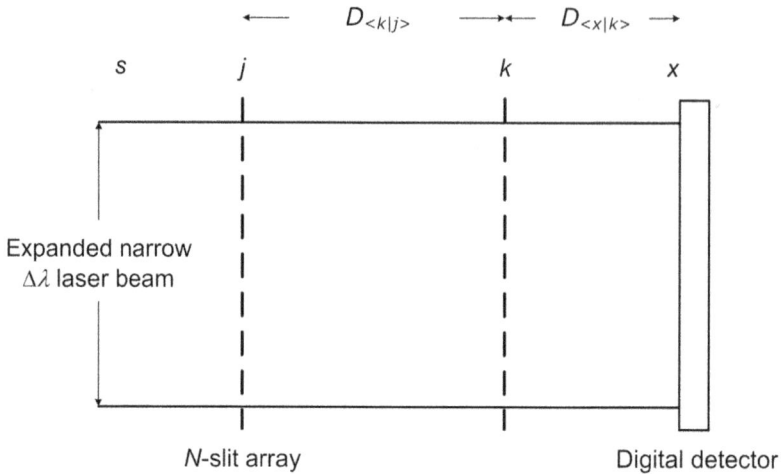

Figure 26.3. An extra plane k is introduced in the intra–interferometric space, severely altering the mathematical nature of the probability amplitude. Now there are two intra–interferometric distances: $D_{\langle x|k \rangle}$ and $D_{\langle k|j \rangle}$.

$$\langle x|s \rangle = \sum_{j=1}^{N} \langle x|k \rangle \langle k|j \rangle \langle j|s \rangle. \tag{26.11}$$

Certainly, these transformations to the probability amplitude transform the probability and ultimately the measured spatial intensity that depends directly on the probability distribution.

The effect of inserting very thin high-surface-quality beam splitters in an attempt to extract information from the propagating interferometric character demolishes the propagating interferogram and has been documented amply in the literature (Duarte 2002, 2005, 2016).

Even attempts to utilize extremely subtle *non-demolition* methods of interception, via the use of microscopic transparent spider web silk fibers, transform the propagating interferogram, as documented by Duarte *et al* (2013). These authors documented the very first diffraction pattern superimposed on one of the 'wings' of a propagating interferogram. The interferogram corresponds to the IQC b ($N = 3$), and is illustrated in its original form in figure 26.4. The intercepted IQC b obtained by measurement is depicted in figure 26.5 while the corresponding intercepted IQC predicted by theory is shown in figure 26.6 (Duarte *et al* 2013).

For readers interested in images of interferograms and IQCs generated experimentally and via equations (26.6) and (26.7), the following publications are recommended: Duarte (1993, 2002, 1995, 2014, 2015). For readers interested in observing the catastrophic collapse of IQCs due to attempts of interception using beam splitters, the following publications are recommended: Duarte (2002, 2005, 2014, 2015, 2016).

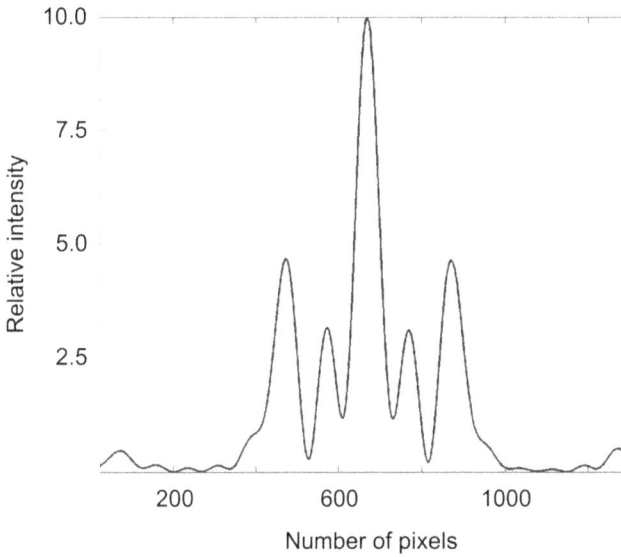

Figure 26.4. Control interferometric character b, corresponding to $N = 3$, at $D_{\langle x|j \rangle} = 7.235$ m and $\lambda = 632.82$ nm. This interferometric character is perfectly represented by the probability amplitude given in equation (26.2). Each pixel is 25 μm wide (from Duarte *et al* 2013).

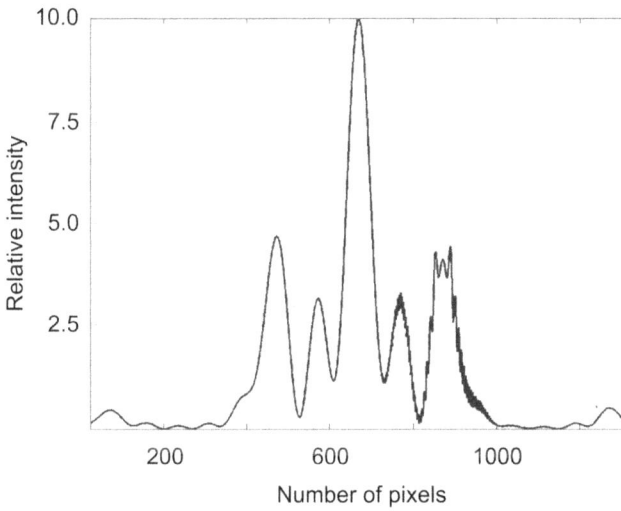

Figure 26.5. Interferometric character b, corresponding to $N = 3$, being intercepted by a transparent spider web silk fiber with a 25 μm diameter at $D_{\langle k|j \rangle} = 7.235 - 0.150$ m and $D_{\langle x|k \rangle} = 0.150$ m, with $\lambda = 632.82$ nm. Even with this extremely subtle interception mechanism, equation (26.2) ceases to describe the measurement. Each pixel is 25 μm wide (Duarte *et al* 2013).

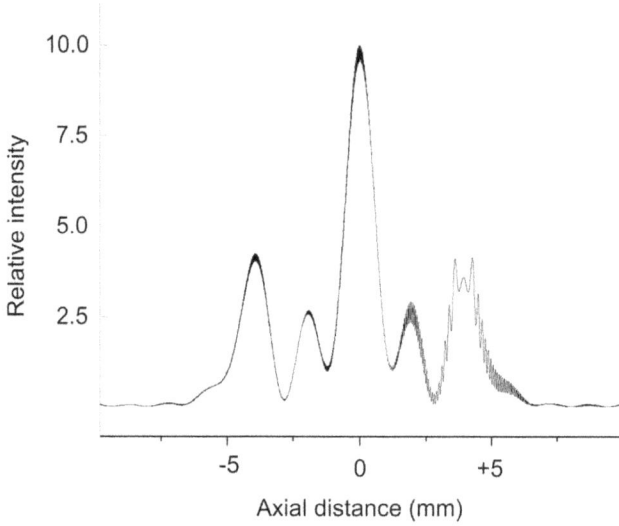

Figure 26.6. Theoretical prediction of the interferometric character b, corresponding to $N = 3$, being intercepted by a transparent spider web silk fiber with a 25 μm diameter at $D_{\langle k|j \rangle} = 7.235 - 0.150$ m, and $D_{\langle x|k \rangle} = 0.150$ m, $\lambda = 632.82$ nm (Duarte *et al* 2013).

26.5 Discussion

Previously, in chapter 17, it was shown that the probability amplitude for quantum entanglement

$$|\psi\rangle = \frac{1}{\sqrt{2}}(|x\rangle_1 |y\rangle_2 - |y\rangle_1 |x\rangle_2)$$

can be derived from the Diracian probability amplitude

$$\langle x|s\rangle = \sum_{j=1}^{N} \langle x|j\rangle\langle j|s\rangle$$

and directly applied to N-slit interferometry (Duarte 2013a, 2013b, 2014). This immediately implies that the core physics for phenomenology derived from either equation is subject to identical quantum principles (Duarte and Taylor 2015). This means that the transmission of information, or communications, based on either equation is intrinsically shielded from in-transit observation.

From a technological–engineering perspective, communications based on quantum entanglement physics appear to be more suitable to long-range communications. However, communications based on QICs generated via N-slit quantum interferometry *do not require* quantum key distribution, offer very attractive signal-to-noise ratios, and are based on a relatively simple optical configuration.

These qualities are technologically attractive for satellite arrays in LEO requiring links in the 10^3–10^4 km range.

References

Bell J S 1964 On the Einstein–Podolsky–Rosen paradox *Physics* **1** 195–200

Born M 1926 Zur quantenmechanik der stoßvorgänge *Z. Phys.* **37** 863–7

Dirac P A M 1939 A new notation for quantum mechanics *Math. Proc. Camb. Philos. Soc.* **35** 416–8

Dirac P A M 1978 *The Principles of Quantum Mechanics* 4th edn (Oxford: Oxford University Press)

Duarte F J 1991 Dispersive dye lasers *High Power Dye Lasers* ed F J Duarte (Berlin: Springer) ch 2

Duarte F J 1993 On a generalized interference equation and interferometric measurements *Opt. Commun.* **103** 8–14

Duarte F J 1995 Interferometric imaging *Tunable Laser Applications* ed F J Duarte (New York: Marcel Dekker) ch 5

Duarte F J 2002 Secure interferometric communications in free space *Opt. Commun.* **205** 313–9

Duarte F J 2005 Secure interferometric communications in free space: enhanced sensitivity for propagation in the metre range *J. Opt. A: Pure Appl. Opt.* **7** 73–5

Duarte F J 2013a The probability amplitude for entangled polarizations: an interferometric approach *J. Mod. Opt.* **60** 1585–7

Duarte F J 2013b Tunable laser optics: applications to optics and quantum optics *Prog. Quantum Electron.* **37** 326–47

Duarte F J 2014 *Quantum Optics for Engineers* (New York: CRC)

Duarte F J 2015 *Tunable Laser Optics* 2nd edn. (New York: CRC)

Duarte F J 2016 Secure space-to-space interferometric communications and its nexus to the physics of quantum entanglement *Appl. Phys. Rev.* **3** 041301

Duarte F J and Paine D J 1989 Quantum mechanical description of N-slit interference phenomena *Proc. of the Int. Conf. on Lasers '88* ed R C Sze and F J Duarte (McLean, VA: STS Press) pp 552–4

Duarte F J and Piper J A 1982 Dispersion theory of multiple-prism beam expanders for pulsed dye lasers *Opt. Commun.* **43** 303–7

Duarte F J, Taylor T S, Black A M and Olivares I E 2013 Diffractive patters superimposed over propagating *N*-slit interferograms *J. Mod. Opt.* **60** 136–40

Duarte F J, Taylor T S, Clark A M and Davenport W E 2010 The *N*-slit interferometer: an extended configuration *J. Opt.* **12** 015705

Duarte F J, Taylor T S, Clark A M, Davenport W E and Varmette P G 2011 *N*-slit interferometer for secure free-space optical communications: 525 m intra interferometric path length *J. Opt.* **13** 035710

Duarte F J and Taylor T S 2015 Quantum entanglement physics secures space-to-space interferometric communications *Laser Focus World* **51** 54–8

Feynman R P, Leighton R B and Sands M 1965 *The Feynman Lectures on Physics* vol III (Reading, MA: Addison-Wesley)

Pryce M L H and Ward J C 1947 Angular correlation effects with annihilation radiation *Nature* **160** 435

Ward J C 1949 *Some Properties of the Elementary Particles* (Oxford: Oxford University Press)

IOP Publishing

Fundamentals of Quantum Entanglement

F J Duarte

Chapter 27

Quanta pair sources for quantum entanglement experiments

The principles of the various sources applied to generate polarization-entangled quanta pairs are briefly reviewed. These include the $e^+e^- \to \gamma_1\gamma_2$ ^{64}Cu sources used in the Compton scattering experiments, the optically pumped ^{40}Ca source used in visible spectrum experiments, and type I and type II SPDC. Possible future sources are also considered.

27.1 Introduction

In this chapter a brief description of the important features of various sources used to generate quanta pairs, exhibiting orthogonal polarizations relative to each other, is provided.

27.2 Positron–electron annihilation

Positron–electron annihilation of the form

$$e^+e^- \to \gamma_1\gamma_2 \tag{27.1}$$

was utilized by Bleuler and Bradt (1948), Hanna (1948), and Wu and Shaknov (1950) to generate gamma rays from ^{64}Cu at a wavelength compatible with an energy of ~1 MeV (Kasday *et al* 1975). This energy translates into a wavelength of $\lambda \approx 1.24 \times 10^{-3}$ nm. However, an additional and specific estimate of the gamma ray wavelength originating from annihilation radiation in ^{64}Cu indicates that the corresponding wavelength is $\lambda \approx 2.42 \times 10^{-3}$ nm (Dumond *et al* 1949).

    The three original experiments utilized deuteron bombardment to activate the ^{64}Cu source, which was confined at the center of a narrow cavity within a solid block of lead, as illustrated in figure 27.1. The geometrical disposition of the ^{64}Cu emitter within the cavity restricts emission in the $+z$ and $-z$ directions, which is one of the central requirements of the original theory (Dirac 1930, Wheeler 1946, Pryce and

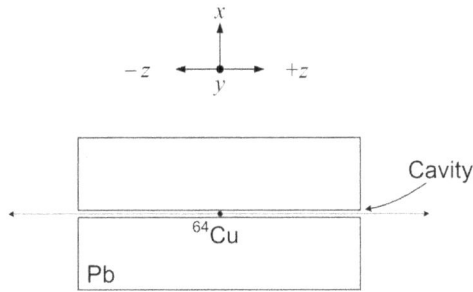

Figure 27.1. Generic schematics for the $e^+e^- \rightarrow \gamma_1\gamma_2$ experiments. The annihilation radiation occurs in all directions but is confined in the $+z$ and $-z$ directions by the cavity, which can have a diameter of $\sim$0.5 mm.

Ward 1947, Ward 1949). It is reasonable to assume that the $+z$ and $-z$ emission, originally labeled as $+k$ and $-k$, is within the $\lambda \pm \Delta\lambda$ distribution.

The correlation features of this emission are the following:

1. Both quanta are emitted collinearly in the $+z$ and $-z$ directions to satisfy the pair theory, which is articulated by Wheeler's dictum: 'if one of these photons is linearly polarized in one plane, then the photon that goes off in the opposite direction with equal momentum is linearly polarized in the perpendicular plane' (Wheeler 1946).

2. The emitted quanta have correlated polarizations as explained above. In this annihilation emission the conservation of momenta and the emission of quanta with orthogonal polarizations are inherently related.

3. Originating from the same ^{64}Cu decay transition, both quanta are assumed to have the same frequencies ($\nu_1 = \nu_2$) and are therefore indistinguishable in the frequency domain.

4. Both quanta are emitted from the same active volume.

27.3 Atomic Ca emission

The use of ^{40}Ca as an active medium was introduced by Kocher and Commins (1967). These authors used a H_2 arc lamp for UV excitation. This optical pumping leads to the population of an upper-lying level in ^{40}Ca that results in the emission of two-step sequential transitions $4p^2\,{}^1S_0 - 4p4s\,{}^1P_1$ ($\lambda_1 = 551.3$ nm) and $4p4s\,{}^1P_1 - 4s^2\,{}^1S_0$ ($\lambda_2 = 422.7$ nm) with correlated polarizations; see figure 27.2.

Aspect *et al* (1981) replaced the UV arc lamp with a scheme using two lasers to cover the excitation transition $4s^2\,{}^1S_0 - 4p^2\,{}^1S_0$. This two-photon excitation, outlined in figure 27.3, utilized a single-mode krypton ion laser emitting at $\lambda_{e1} = 406.7$ nm and a tunable dye laser emitting at $\lambda_{e2} = 581$ nm. The two excitation lasers are said to have 'parallel polarizations' and the emission photons ν_1 and ν_2 have 'correlated polarizations' (Aspect *et al* 1981).

The correlation features of this emission are the following:

1. The quanta pair is emitted collinearly in the $+z$ and $-z$ directions.

2. The emission of both quanta originates from the same $4s^2\,{}^1S_0$ upper level of ^{40}Ca.

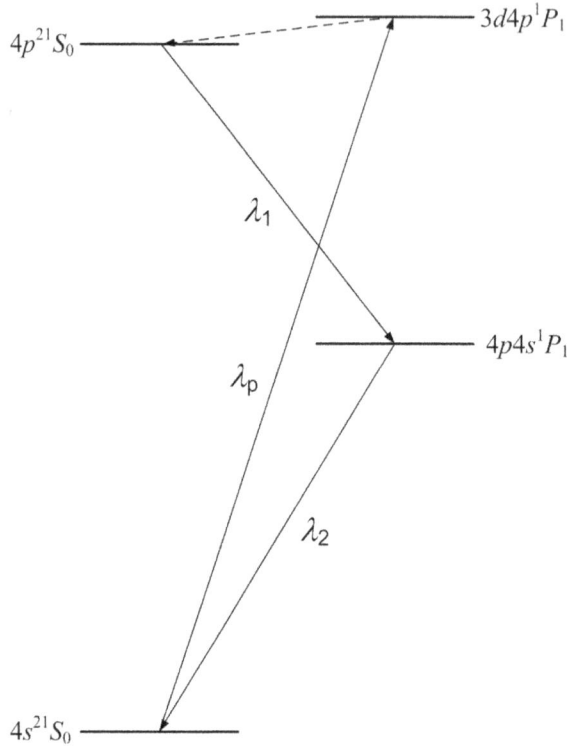

Figure 27.2. Energy level diagram for ^{40}Ca relevant to UV arc lamp excitation. The cascade emission with correlated polarizations involve the $4p^{21}S_0 - 4p4s^1P_1$ ($\lambda_1 = 551.3$ nm) and $4p4s^1P_1 - 4s^{21}S_0$ ($\lambda_2 = 422.7$ nm) transitions.

3. The emitted quanta, ν_1 and ν_2, have correlated polarizations.
4. Both quanta are emitted from the same active volume.

However, it is necessary to note that the emission quanta wavelengths, $\lambda_1 = 551.3$ nm $\lambda_2 = 422.7$ nm, do not lead to indistinguishability or $|p_1| = |p_2|$. This is a departure from the original requirement introduced by Wheeler (1946).

27.4 Type I SPDC

Type I SPDC as a source of pairs of entangled photons was demonstrated by Shih and Alley (1988). This approach takes advantage of the phase matching conditions

$$\omega = \omega_1 + \omega_2 \tag{27.2}$$

$$\hat{k} = \hat{k}_1 + \hat{k}_2 \tag{27.3}$$

where ω is the frequency of the pump laser while ω_1 and ω_2 are the frequencies of the two quanta, emitted in divergent directions, from the nonlinear crystal. Similarly, $\hat{k}$

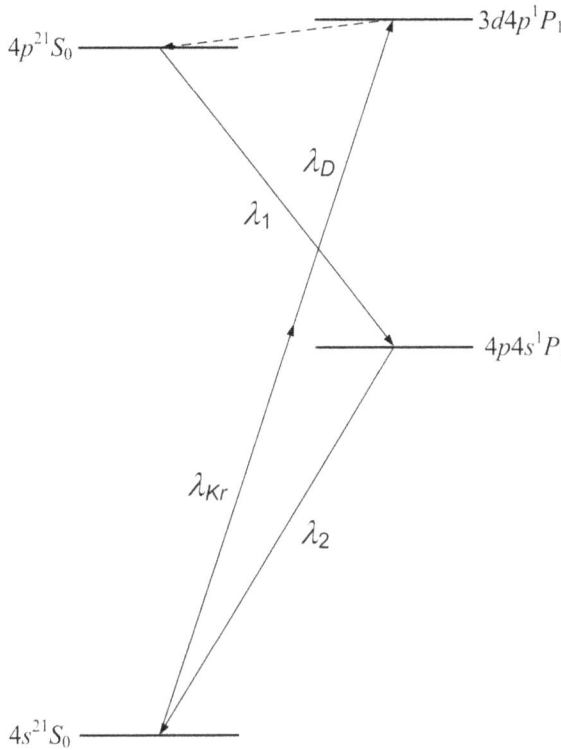

Figure 27.3. Approximate energy level diagram for ^{40}Ca relevant to the Aspect experiments. Aspect *et al* (1981) replaced the UV arc lamp with two counter propagating lasers at $\lambda_{e1} = 406.7$ nm and $\lambda_{e2} = 581$ nm. The cascade emission with correlated polarizations involves the $4p^{21}S_0 - 4p4s^1P_1$ ($\lambda_1 = 551.3$ nm) and $4p4s^1P_1 - 4s^{21}S_0$ ($\lambda_2 = 422.7$ nm) transitions.

is the wave vector of the pump laser while $\hat{k}_1$ and $\hat{k}_2$ are the wave vectors of the two quanta emitted in divergent directions (see figure 27.4(a)). For an extensive review on type I parametric down-conversion, the reader is referred to Orr *et al* (2016).

In their work, Shih and Alley (1988) utilized the fourth harmonic of a Nd:YAG laser, at $\lambda \approx 266$ nm, as the excitation source for a KD* P crystal. The emission was at $\lambda \approx 532$ nm. Since $\hat{k}$, $\hat{k}_1$, and $\hat{k}_2$ must be in the same plane, coplanar irises were employed as spatial filters for ω_1 and ω_2. The emission for ω_1 and ω_2 was inherently polarized in the $|y\rangle$ state. Following each iris, a $\lambda/4$ wave plate was deployed to produce two left-hand circularly polarized beams $|L\rangle$; one of those beams underwent two further reflections to produce a right-hand circularly polarized beam $|R\rangle$. Thus, the entanglement was examined using pairs of photons in the $|L\rangle$ and $|R\rangle$ states (Shih and Alley 1988).

It should be pointed out that with ω_1 and ω_2 polarized in the $|y\rangle$ state, the simple deployment of a $\pi/2$ polarization rotator in one of the beams should produce the desired $|x\rangle$ and $|y\rangle$ states.

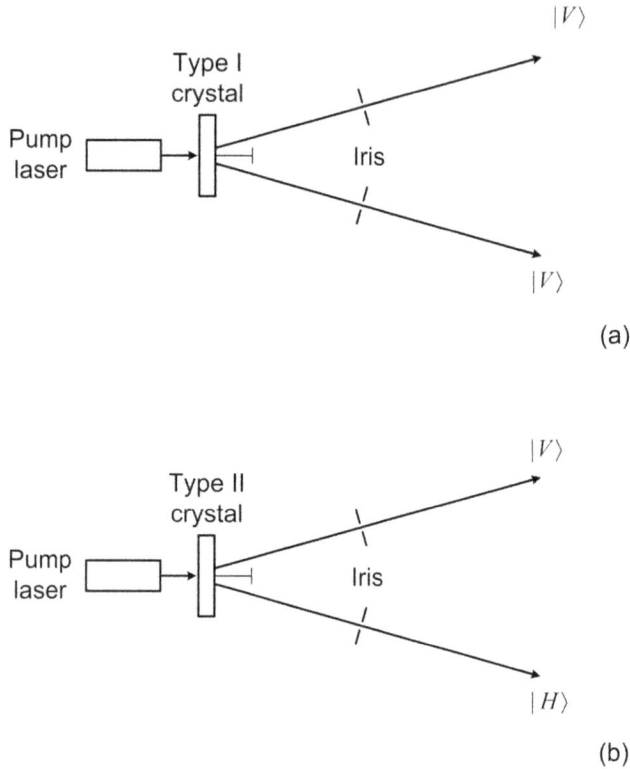

Figure 27.4. Generic diagram applicable to (a) type I SPDC and (b) type II SPDC. In both cases irises are used to select coplanar emission (see text).

The correlation features of this emission are the following:
1. The emitted quanta, ν_1 and ν_2, have correlated identical polarizations. This feature requires the use of external optics to obtain $|L\rangle$ and $|R\rangle$ states or $|x\rangle$ and $|y\rangle$ states.
2. The emitted quanta have the same frequencies ($\nu_1 = \nu_2$) and are, therefore, indistinguishable in the frequency domain.
3. The emission of both quanta originates from the same active volume.

27.5 Type II SPDC

Inherently entangled $|x\rangle$ and $|y\rangle$ states, or $|H\rangle$ and $|V\rangle$ states, can be produced from type II SPDC as reported by Kwiat *et al* (1995). In type II SPDC the orthogonal states of polarization are generated from a single nonlinear crystal via ordinary and extraordinary polarization. The emission axes are at an angle to each other and the emission occurs distributed within two propagation cones. Conical refraction in crystals is described by Born and Wolf (1999).

According to Kwiat *et al* (1995), the emission of ω_1 and ω_2 is inherently entangled according to

$$|\psi\rangle = \frac{1}{\sqrt{2}}(|H\rangle_1\,|V\rangle_2 + e^{i\alpha}\,|V\rangle_1\,|H\rangle_2) \tag{27.4}$$

and via the appropriate use of $\lambda/2$ and $\lambda/4$ wave plates to control the value of α on one of the emission beams, the complete set of entangled states

$$|\psi\rangle = \frac{1}{\sqrt{2}}(|H\rangle_1\,|V\rangle_2 \pm |V\rangle_1\,|H\rangle_2) \tag{27.5}$$

$$|\psi\rangle = \frac{1}{\sqrt{2}}(|H\rangle_1\,|H\rangle_2 \pm |V\rangle_1\,|V\rangle_2) \tag{27.6}$$

can be produced. These authors used argon ion laser excitation ($\lambda = 351.1$ nm) of a β-barium borate crystal emitting at $\lambda \approx 702$ nm, and the ω_1 and ω_2 photons were guided toward their respective detectors via coplanar irises as in the type I arrangement of Shih and Alley (1988); see figure 27.4(b).

In an alternative experimental arrangement, Yin *et al* (2017) reported on the use of a PPKTP crystal within a Sagnac interferometer. The pump laser was a 30 mW diode laser emitting at $\lambda \approx 405$ nm with a linewidth of $\Delta\nu = 160$ MHz. The emission was reported as 5.9×10^6 photon pairs per second with a fidelity of 0.907 ± 007 at $\lambda \approx 810$ nm. A description of the Sagnac interferometer is given in appendix D.

The correlation features of this emission are the following:

1. The emitted quanta, ν_1 and ν_2, have correlated polarizations with the polarization of one quanta $|x\rangle$ perpendicular to the polarization of the other quanta $|y\rangle$.
2. The emitted quanta have the same frequencies ($\nu_1 = \nu_2$) and are therefore indistinguishable in the frequency domain.
3. The emission of both quanta originates from the same active volume.

27.6 Further horizons

Given its solid-state status and its compactness, SPDC has become by far the most pervasive avenue to achieve emission of polarization-entangled photon pairs. Limitations of this methodology include the use of an indirect method of excitation, via optical pumping, and very low conversion efficiencies as becomes apparent from the figures provided by Yin *et al* (2017): only 5.9×10^6 photon pairs per second were generated from a narrow-linewidth diode laser excitation of 30 mW at $\lambda \approx 405$ nm.

One area of alternative emission device development that has received attention is quantum dot *biexiton* sources. For instance, an InAs quantum dot embedded in a p–i–n heterostructure planar microcavity was reported to yield circularly polarized photons in the state

$$|\psi\rangle = \frac{1}{\sqrt{2}}(|R\rangle\,|L\rangle + |L\rangle\,|R\rangle) \tag{27.7}$$

with a fidelity of $f = 0.826 \pm 0.027$ at ~1.4 eV, or $\lambda \approx 885$ nm (Salter *et al* 2010). The InAs quantum dot layer was configured at the center of a GaAs microcavity.

These authors did not provide information on the output number of photons or emission intensity. Müller *et al* (2014) reported on a biexiton device yielding linearly polarized entangled photon pair generation in the state

$$|\psi\rangle = \frac{1}{\sqrt{2}}(|H\rangle |H\rangle + |V\rangle |V\rangle) \tag{27.8}$$

with a fidelity of $f = 0.81 \pm 0.02$ at a visibility of $V = 0.86 \pm 0.03$. The emission took place at ~1.42 eV, or $\lambda \approx 873$ nm (Müller *et al* 2014).

Huber *et al* (2017) reported on a GaAs quantum dot biexiton device yielding a photon pair generation in the state

$$|\psi\rangle = \frac{1}{\sqrt{2}}(|H\rangle |H\rangle + e^{i\alpha\pi} |V\rangle |V\rangle) \tag{27.9}$$

with $\alpha = 0.12$. The reported fidelity was $f = 0.94 \pm 0.01$ at a visibility of $V = 0.93 \pm 0.07$. This was a pulsed device excited every 12.5 ns emitting at ~1.57 eV ($\lambda \approx 790$ nm). Again, no information on the net number of entangled photons was given. However, it is indicated that the biexiton emission yields, on average, about 1000 counts per second. This type of photon count per second, from semiconductor emitters, can be consistent with peak powers in the nanowatt range (Duarte *et al* 2005). Efforts to generate indistinguishable photons from multiple quantum dots have also been reported (Singh *et al* 2019).

A further possible alternative is to learn to apply coherent sources, such as modified tunable narrow-linewidth lasers, that yield *populations of indistinguishable photons*. An incipient discussion in this direction has been recently provided (Duarte 2018).

References

Aspect A, Grangier P and Roger G 1981 Experimental tests of realistic local theories via Bell's theorem *Phys. Rev. Lett.* **47** 460–3

Bleuler E and Bradt H L 1948 Correlation between the states of polarization of the two quanta of annihilation radiation *Phys. Rev.* **73** 1398

Born M and Wolf E 1999 *Principles of Optics* 7th edn (Cambridge: Cambridge University Press)

Dirac P A M 1930 On the annihilation of electrons and protons *Math. Proc. Camb. Philos. Soc.* **2** 361–75

Duarte F J 2018 Organic lasers for *N*-channel quantum entanglement *Organic Lasers and Organic Photonics* ed F J Duarte (Bristol: Institute of Physics) ch 15

Duarte F J, Liao L S and Vaeth K M 2005 Coherence characteristics of electrically excited tandem organic light-emitting diodes *Opt. Lett.* **30** 3071–4

Dumond J W M, Lind D A and Watson B B 1949 Precision measurement of the wavelength and spectral profile of the annihilation radiation from Cu^{64} with the two-meter focusing curved crystal spectrometer *Phys. Rev.* **75** 1226–39

Hanna R C 1948 Polarization of annihilation radiation *Nature* **162** 332

Huber D, Reindl M, Huo Y, Huang H, Wildmann J S, Schmidt O G, Rastelli A and Trotta R 2017 Highly indistinguishable and strongly entangled photons from symmetric GaAs quantum dots *Nat. Commun.* **8** 15506

Kasday L R, Ullman J D and Wu C S 1975 Angular correlation of Compton-scattered annihilation photons and hidden variables *Il Nuovo Cimento* **25** 633–61

Kocher C A and Commins E D 1967 Polarization correlation of photons emitted in an atomic cascade *Phys. Rev. Lett.* **18** 575–7

Kwiat P G, Mattle K, Weinfurter H and Zeilenger A 1995 New high-intensity source of polarization-entangled photon pairs *Phys. Rev. Lett.* **75** 4337–41

Müller M, Bounouar S, Jöns K D and Michler P 2014 On-demand generation of indistinguishable polarization-entangled photon pairs *Nat. Photon.* **8** 224–8

Orr B J, Haub J G, He Y and White R T 2016 Spectroscopic applications of tunable optical parametric oscillators *Tunable Laser Applications* (New York: CRC) ch 2

Pryce M L H and Ward J C 1947 Angular correlation effects with annihilation radiation *Nature* **160** 435

Salter C L, Stevenson R M, Farrer I, Nicoll C A, Ritchie D A and Shields A J 2010 An entangled-light-emitting diode *Nature* **465** 594–7

Shih Y H and Alley C O 1988 New type of Einstein–Podolsky–Rosen–Bohm experiment using pairs of light quanta produced by optical parametric down conversion *Phys. Rev. Lett.* **61** 2921–4

Singh A *et al* 2019 Quantum frequency conversion of a quantum dot single-photon source on a nanophotonic chip *Optica* **6** 563–9

Ward J C 1949 *Some Properties of the Elementary Particles* (Oxford: Oxford University Press)

Wheeler J A 1946 Polyelectrons *Ann. N. Y. Acad. Sci.* **48** 219–38

Wu C S and Shaknov I 1950 The angular correlation of scattered annihilation radiation *Phys. Rev.* **77** 136

Yin J *et al* 2017 Satellite-based entanglement distribution over 1200 kilometers *Science* **356** 1140–4

Chapter 28

More on quantum entanglement

Various aspects of the physics of quantum entanglement are discussed further. For instance, the impervious nature of quantum mechanics relative to hidden variable theory arguments. Moreover, the importance of indistinguishability in the approach of Dirac to his quantum state identities is underlined. Photon non-locality is emphasized.

28.1 Introduction

In this chapter the subject of quantum entanglement is revisited via various concepts previously considered only tangentially or not considered at all.

28.2 Consequences of the EPR paper

The argument on the 'incompleteness' of quantum mechanics put forward by Einstein *et al* (1935) initiated the philosophical path toward quantum entanglement and had the unintended consequence of ultimately bringing about Bell's theorem (Bell 1964), which reinforced the incompatibility of 'local hidden variable theories' with quantum mechanics and is today widely applied as a tool to verify the integrity of quantum cryptography.

However, as explained in appendix A, the EPR argument of incompleteness of quantum mechanics can be neutralized via Heisenberg's uncertainty principle. Knowing that there is a physical, non-philosophical, path toward quantum entanglement, this observation might explain why Willis Lamb classified the EPR paradox as a 'misunderstanding' and Bell's theorem, or Bell's inequalities, as 'unnecessary' (Lamb 2001).

28.3 Hidden variable theories

Hidden variable theories began to surface apparently, in a shadowy unspecified form, in the late 1920s or early 1930s. This is known with a fair degree of certainty because John von Neumann does refer to them explicitly in his book: 'the "hidden parameter" … has been proposed more than once' (von Neumann 1932). These hidden variable theories assume 'dispersion free states' (von Neumann 1932, Bell 1966). In this regard, von Newman is adamant: 'dispersion free ensembles, which would have to correspond to actual states … do not exist' (von Neumann 1932). In fact, from a pragmatic perspective it can be observed that the concept of 'dispersion free states' is incompatible with Heisenberg's uncertainty principle. Nevertheless, the interest in 'dispersion free states' resurfaced with renewed vigor in the 1950s (Bohm 1952, Bohm and Bub 1966).

At this stage it should be made clear that hidden variable theories—'"hidden" because … we can only conjecture their existence and certainly cannot control them' (Bell 1971)—have not led to any practical physics and their only application appears to have been their questionable use as a platform to project doubts toward legitimate quantum entanglement experiments utilizing positron–electron annihilation sources, that is, the experiments of Bleuler and Bradt (1948), Hanna (1948) and Wu and Shaknov (1950). Here, it is relevant to mention that Feynman dismisses the concept of 'inner variables', using the sentence 'this is the way nature really *is*' while discussing quantum interference (Feynman *et al* 1965). Willis Lamb (2001) also called hidden variable theories 'unnecessary'.

Relevant to this argument is also the statement, 'the two photons are entangled and according to local realism, their polarizations planes should become independent … a typical EPR situation. Already in 1948, observations agreed with quantum mechanics … not with local realism' (Dalitz and Duarte 2000).

28.4 The perspectives of EPR and Schrödinger on quantum entanglement

It was EPR (Einstein *et al* 1935) that inspired Schrödinger (1935, 1936) to write his two papers on quantum entanglement. In this regard, EPR argued that 'making predictions concerning a system on the basis of measurements made on another system' was a 'problem' (Einstein *et al* 1935).

In his first paper on quantum entanglement, Schrödinger (1935) wrote, 'after reestablishing one ψ-function … the other one can be inferred simultaneously', and followed up with, 'It is rather discomforting that the theory should allow a system to be steered or piloted into one or the other type of state at the experimenter's mercy in spite of his having no access to it'.

The first part of his analysis is correct since by following a polarization measurement on photon 1, the experimenter *can infer* the polarization state of photon 2. However, this is strictly a statistical outcome since prior to the measurement the experimenter has no information or control on the polarization state of photon 1. The second part of Schrödinger's argument as to 'steering or piloting' the polarization state of photon 2 is not supported by experiment.

28.5 Indistinguishability and Dirac's identities

As highlighted in chapter 17 and previous publications (Duarte 2013a, 2013b, 2014), crucial to the development of the quantum entanglement probability amplitude is the Dirac identity (Dirac 1978)

$$|X\rangle = |a\rangle_1 \, |b\rangle_2 \, |c\rangle_3 \ldots |g\rangle_n \qquad (28.1)$$

which means that indistinguishable quanta 1, 2, 3, ... n can be in different states. When referring to this 'curious phenomena ... having no analogue in classical theory' Dirac is writing about quanta of the 'same kind' and 'absolutely indistinguishable from one another'. At the same time he is also contemplating arrays of *identical* quanta in different states such as $|a\rangle_1$, $|b\rangle_1$, $|c\rangle_1$... and $|a\rangle_2$, $|b\rangle_2$, $|c\rangle_2$.... It is then natural to extend one of Dirac's initial identities to

$$|Y\rangle = |a\rangle_1 \, |b\rangle_1 \, |c\rangle_1 \ldots |g\rangle_1 \qquad (28.2)$$

and

$$|Z\rangle = |a\rangle_2 \, |b\rangle_2 \, |c\rangle_2 \ldots |g\rangle_2 \qquad (28.3)$$

and so on, which apply to the very same, utterly indistinguishable, quanta in different states. This leads to the concept of a series of indistinguishable quanta in different states of polarization (see chapter 20). More specifically, these Dirac identities are applicable, for instance, to a single-transverse-mode single-longitudinal-mode laser beam of indistinguishable quanta including both $|x\rangle$ and $|y\rangle$ states of polarization. This has not yet been observed experimentally.

28.6 Photon non-locality

In chapter 1 it was mentioned that from Heisenberg's uncertainty principle

$$\Delta p \, \Delta x \approx h \qquad (28.4)$$

the space–frequency identity

$$\Delta \nu \, \Delta x \approx c \qquad (28.5)$$

can be derived (Duarte 2003). This is quite real to any observant experimental physicist that has played with Mach–Zehnder interferometers. If such an interferometer is built with a narrow-linewidth laser exhibiting a linewidth of $\Delta \nu$, then the overall linear space of intra–interferometric distance for which interference will be observed is

$$\Delta x \approx \frac{c}{\Delta \nu}. \qquad (28.6)$$

Single photons exhibit extremely narrow linewidths and correspondingly can exhibit *enormous* coherence lengths.

In a quantum entanglement experiment, two indistinguishable, extremely narrow-linewidth photons are emitted in opposite directions. This extremely narrow linewidth allows an extremely large coherence length determined by the analogous expression of the uncertainty principle $\Delta x \approx c/\Delta \nu$. Hence, it is not difficult to observe that the emitted quanta will continue to interact after emission from their common source. The implications of photon non-locality have yet to be fully explored experimentally.

28.7 Discussion

The material in this chapter reaffirms the enormous difference between the way of thinking about the same physical issue by Einstein and Schrödinger, on the one hand, and Dirac and Wheeler, on the other hand. Certainly, Dirac was not bothered at all by issues of pre-established philosophy. The fact that local hidden variable theories led to *nowhere* is a strong reaffirmation of the success of the pragmatic way of thinking implicitly adopted by Dirac.

The Dirac identities $|X\rangle = |a\rangle_1 |b\rangle_2 |c\rangle_3 \ldots |g\rangle_n$, $|Y\rangle = |a\rangle_1 |b\rangle_1 |c\rangle_1 \ldots |g\rangle_1$, and $|Z\rangle = |a\rangle_2 |b\rangle_2 |c\rangle_2 \ldots |g\rangle_2$ convey profound teachings that have yet to be fully harvested.

References

Bell J S 1964 On the Einstein–Podolsky–Rosen paradox *Physics* **1** 195–200

Bell J S 1966 On the problem of hidden variables in quantum mechanics *Rev. Mod. Phys.* **38** 447–52

Bell J S 1971 Introduction to the hidden variable question *Proc. of the Int. School of Physics Enrico Fermi: Foundations of Quantum Mechanics* (New York: Academic), pp 171–81

Bleuler E and Bradt H L 1948 Correlation between the states of polarization of the two quanta of annihilation radiation *Phys. Rev.* **73** 1398

Bohm D 1952 A suggested interpretation of quantum theory using 'hidden' variables. I *Phys. Rev.* **85** 166–79

Bohm D and Bub J 1966 A proposed solution of the measurement problem in quantum mechanics by hidden variable theory *Rev. Mod. Phys.* **38** 453–69

Dalitz R H and Duarte F J 2000 John Clive Ward *Phys. Today* **53** 99–100

Dirac P A M 1978 *The Principles of Quantum Mechanics* 4th edn (Oxford: Oxford University Press)

Duarte F J 2003 *Tunable Laser Optics* (New York: Elsevier)

Duarte F J 2013a The probability amplitude for entangled polarizations: an interferometric approach *J. Mod. Opt.* **60** 1585–7

Duarte F J 2013b Tunable laser optics: applications to optics and quantum optics *Prog. Quantum Electron.* **37** 326–47

Duarte F J 2014 *Quantum Optics for Engineers* (New York: CRC)

Einstein A, Podolsky B and Rosen N 1935 Can quantum mechanical description of physical reality be considered complete? *Phys. Rev.* **47** 777–80

Feynman R P, Leighton R B and Sands M 1965 *The Feynman Lectures on Physics* vol III (Reading, MA: Addison-Wesley)

Hanna R C 1948 Polarization of annihilation radiation *Nature* **162** 332

Lamb W E 2001 Super classical quantum mechanics: the best interpretation of nonrelativistic quantum mechanics *Am. J. Phys.* **69** 413–21

Schrödinger E 1935 Discussion of probability relations between separated systems *Math. Proc. Camb. Philos. Soc.* **31** 555–63

Schrödinger E 1936 Probability relations between separated systems *Math. Proc. Camb. Philos. Soc.* **32** 446–52

von Neumann J 1932 *Mathematische Grundlagen der Quanten-Mechanik* (Berlin: Springer)

Wu C S and Shaknov I 1950 The angular correlation of scattered annihilation radiation *Phys. Rev.* **77** 136

Chapter 29

On the interpretation of quantum mechanics

A pragmatic approach to the interpretation of quantum mechanics, unbound by pre-existing philosophical prejudices, is advocated following the approach and attitude to this subject by luminaries such as Dirac, Feynman, Lamb, and Ward. The crucial importance of the Dirac–Feynman probability amplitude is emphasized while the suggestion of this being a quantum world, as hinted by Lamb, is adopted.

29.1 Introduction

Much has been written on the interpretation of quantum mechanics. Besides the original orthodox interpretation known as the Copenhagen interpretation, outlined mainly by Bohr and Heisenberg, there exists a long series of additional interpretations. A perspective on the Copenhagen interpretation is given by Bohr using the principle of *complementarity*, which rests on Heisenberg's uncertainty principle, while responding to the arguments of Einstein *et al* (1935) (Bohr 1935). In this regard, it is useful to quote from another of the quantum originals: 'the laws of quantum mechanics were found by the slow and tedious process of interpreting experimental results' (Born 1948).

It is certainly not the intention of this chapter to provide an overview of this vast subject matter, and the numerous interpretations available, except to exhibit the very basic principles that make quantum mechanics work and serve as a highly practical theoretical framework particularly in the areas of quantum interferometry and quantum entanglement. In this regard, it is certainly relevant to quote from one of Dirac's last papers: 'The interpretation of quantum mechanics has been dealt with by many authors, and I do not want to discuss it here. I want to deal with more fundamental things' (Dirac 1987).

It is also relevant to recall Ward's observation: '$(|x, y\rangle - |y, x\rangle) \ldots$ *was my first lesson in quantum mechanics, and in a very real sense my last, since all the rest is mere technique, which can be learnt from books*' (Ward 2004).

In all fairness it should be mentioned that physicists such as Dirac and Ward are two of those who, according to Bell, had quantum mechanics *'in their bones'* and that in regard to the issue of interpretations would respond … *'Why bother?'* (Bell 1990).

29.2 Quantum critical

Criticism is good and it should be welcomed especially when formulated with the intention of improving the subject matter at the center of the criticism. That appears to have been Bell's intent. Here, two of his papers critical of quantum mechanics are considered while focusing on some specific issues.

In 1990, a cleverly written essay sharply critical of quantum mechanics, and its orthodox interpretation, was published by none other than J S Bell. The article was entitled 'Against "measurement"' (Bell 1990). In it, Bell criticized Dirac for having a *'why bother'* attitude towards the possibility of an 'exact' formulation of the theory.

This was not Bell's only critical piece on quantum mechanics. Previously, Bell had co-authored an article entitled 'The moral aspects of quantum mechanics' in which he concluded that the theory is 'at best, incomplete' and that 'it carries in itself the seeds of its own destruction' (Bell and Nauenberg 1966).

Criticisms of quantum mechanics were nothing new at the time of Bell's writings, indeed as described in chapter 2. The notion of quantum mechanics as an incomplete theory was advanced early on by EPR (Einstein *et al* 1935) and these criticisms were almost immediately seconded by Schrödinger (1935, 1936) in the context of quantum entanglement. These were the criticisms that eventually might have given origin to the fruitless 'hidden variable theories' of Bohm (1952) and Bohm and Bub (1966), although von Neumann had already written in 1932, 'the "hidden param-eter" … has been proposed more than once' (von Neumann 1932); see chapter 28.

The genius of Bell is that he brought a whole new level of expressional cleverness to the criticisms of quantum mechanics. Bell's eloquence and subtlety was clearly recognized in a reply to his 'Against "measurement"' essay by a contemporaneous researcher (Gottfried 1991).

29.2.1 On 'The moral aspects of quantum mechanics'

This co-authored Bell paper is a seven-and-a-half-page-long publication. It deals with the notion of morality in quantum mechanics as related to 'the reduction of the wave packet' (Bell and Nauenberg 1966).

This particular subsection, although it includes the word *moral* in its title, does not deal at all with issues of morality. It only focuses on one aspect of Bell's paper and on its final conclusion that quantum mechanics 'carries the seeds of its own destruction'.

The aspect of this Bell paper that is worthy of attention appears in its sixth page and deals with the double-slit, or Young's, experiment. Here Bell invokes the 'de Broglie–Bohm "pilot wave" or "hidden parameter" interpretation of quantum mechanics' (Bell and Nauenberg 1966). Bell's interpretation of the two-slit experi-ment is not only unnecessary but is at odds with the lucid description given by Dirac 'each photon goes partly into each of the two components' (Dirac 1978), and Feynman's statement that describes the photon as '*neither*' a wave nor a particle (Feynman *et al* 1965). An additional description of the two-slit experiment, as extended to an *N*-slit interferometer, explains that 'all the indistinguishable photons illuminate the array of *N* slits, or grating, simultaneously. If only one photon propagates at any given time, then that individual photon illuminates the whole array of *N* slits simultaneously' (Duarte 2003). It should be added that recent experimental evidence points against the 'pilot wave' interpretation (Andersen *et al* 2015).

Toward the end of page 6, Bell and Nauenberg (1966) arrive at the opinion that quantum mechanics is at 'best incomplete'. As described in chapter 3, the initial claim of incompleteness originated with EPR (Einstein *et al* 1935), and this claim can be classified as incompatible with Heisenberg's uncertainty principle (see appendix A). However, Bell's opinion appears to be based on his own thoughts and premises. One of those thoughts is his understanding of the double-slit experiment via the 'pilot wave interpretation'. However, as discussed previously in this subsection, this interpretation can be classified as either unnecessary or erroneous. In conclusion, from a pragmatic perspective, it appears that quantum mechanics *does not* carry the seeds of its own destruction.

29.2.2 On 'Against measurement'

This solo Bell paper is also a seven-and-a-half-page-long publication. It presents a strong criticism on the use of the word *measurement* in quantum mechanics (Bell 1990). It was Bell's last paper.

Again, the purpose of this subsection is not to criticize the whole essay but to focus on a couple of interesting aspects.

In the second page of this essay, Bell suggests that it would be good to replace the word *measurement* with the word *experiment* (our italics). Then he goes on to say, 'A serious formulation will not exclude the big world outside the laboratory' (Bell 1990). In this regard, it should be indicated that experiments with very large *N*-slit interferometers, utilizing populations of indistinguishable photons and measurement predictions based on Diracian probability amplitudes, have taken place within the laboratory at an intra–interferometric distance of $D_{\langle x|j \rangle} = 35$ m and *outside* the laboratory at $D_{\langle x|j \rangle} = 527$ m (Duarte *et al* 2010, 2011). More recently, space-to-Earth quantum entanglement experiments, utilizing the quantum entanglement probability amplitudes, have been reported over a propagation distance of 1200 km (Yin *et al* 2017).

On the third page, Bell considers the book of Landau and Lifshitz (1977). Although of a more marginal relevance, this criticism is nevertheless included here since it illustrates Bell's critical finesse. Bell criticized Landau and Lifshitz for the statement 'It is in principle impossible … to formulate the basic concepts of quantum mechanics without using classical mechanics' (Landau and Lifshitz 1977). Given the use of the word 'impossible', this is debatable. Firstly, as discussed in chapter 1, quantum mechanics' most basic equation $E = h\nu$ originated from an entirely classical macroscopic experiment. Secondly, as illustrated in chapter 17, the probability amplitude for quantum entanglement is derived entirely from quantum concepts.

According to Bell, the problem with Landau and Lifshitz's formulation of quantum mechanics was their *ambiguity* in regard to 'what is microscopic, what is macroscopic, what [is] quantum, [and] what [is] classical' (Bell 1990). Here it should be indicated that from W E Lamb's perspective, quantum mechanics does apply to large-scale phenomena (see section 29.4). In this regard, it could be argued that Landau and Lifshitz's ambiguity might not be a detrimental issue. In fact, if ambiguity is defined from a context of *uncertainty* then it is not a unique concept in quantum mechanics. Ambiguity is also related to doubt … and actually, Feynman celebrated the value of doubt (Feynman 1998).

29.3 Pragmatic perspective

From a pragmatic perspective, which is the perspective of this monograph, the following observations are relevant:

1. A theory should be criticized if it fails to *reproduce* experimental results. This *is not* the case with quantum mechanics.
2. A theory should be criticized if it fails to *predict* experimental results. This *is not* the case with quantum mechanics.
3. A theory should be criticized if it fails to predict measurable phenomena. This *is not* the case with quantum mechanics.
4. A theory should be criticized if it leads to ugly and messy equations. This *is not* the case with quantum mechanics, at least as presented in this monograph.

It can now be said that quantum mechanics neatly fulfils the definition of a correct theory as outlined by EPR (Einstein *et al* 1935): 'The correctness of the theory is judged by the degree of agreement between the conclusions of the theory and human experience. This experience … in physics takes the form of experiment and measurement' (Einstein *et al* 1935).

29.4 Fundamental principles

From an interferometric perspective, the very basic quantum mechanical framework needed to make the physics work is based on the following laws, and/or principles:

1. Planck's quantum energy equation (Planck 1901)

$$E = h\nu. \tag{29.1}$$

2. Heisenberg's uncertainty principle (Heisenberg 1927)

$$\Delta p \, \Delta x \approx h. \tag{29.2}$$

3. The Dirac principles as outlined by Feynman (Feynman *et al* 1965)

$$\langle j|i \rangle = \delta_{ji} \tag{29.3}$$

$$\langle\chi|\phi\rangle = \sum_{i=1}^{N}\langle\chi|i\rangle\langle i|\phi\rangle \qquad (29.4)$$

$$\langle\chi|\phi\rangle = \langle\phi|\chi\rangle^{*}. \qquad (29.5)$$

4. Dirac's identity for 'similar particles' (Dirac 1978)

$$|X\rangle = |a\rangle_1 |b\rangle_2 |c\rangle_3 \ldots |g\rangle_n. \qquad (29.6)$$

5. Born's rule (Born 1926)

$$|\langle\chi|\phi\rangle|^2 = \langle\chi|\phi\rangle\langle\chi|\phi\rangle^{*}. \qquad (29.7)$$

These seven laws, and/or principles, of quantum mechanics provide the foundations of the knowledge needed to quantify generalized quantum interference and quantum entanglement.

29.5 The Dirac–Feynman–Lamb doctrine

There is a portion of the quantum optics community that considers only single-photon events to be quantum events. This is a very limited, and myopic, perspective that ignores epic historical developments in quantum physics. First of all, it was an experimental event in the *very macroscopic domain* that led Planck to discover the quantum energy equation $E = h\nu$ (Planck 1901) that indeed *was* the discovery of quantum mechanics. Secondly, it was Dirac's discussion on macroscopic two-beam interference, in reference to beams of *populations of indistinguishable photons*, that led Dirac to his famous dictum: *'Each photon then interferes only with itself. Interference between two different photons never occurs'* (Dirac 1978). A dictum that even today leads some into confusion. Thirdly, Feynman selected macroscopic beam divergence to illustrate the application of his quantum path integrals (Feynman and Hibbs 1965). This recitation of quantum facts should be integrated to van Kampen's second theorem on quantum measurements: 'Quantum mechanics is concerned with macroscopic phenomena which are not perturbed by observations' (van Kampen 1988) even though the words *applicable to* rather than *concerned with* would give this theorem a more universal usage.

Finally, the doctrine being introduced here is not complete without the sage Willis Lamb. This 'rare theorist turned experimentalist' (Kaiser 2005) stated categorically, 'in principle we could apply quantum descriptions to large scale phenomena, but, in general, we can't because we are unwilling or unable to adhere to the rules of the game … it is high time that we realize that this is inevitable and learn to enjoy it' (Lamb 1987).

The quantum mechanical description of interference applies equally well to situations involving either single photons or populations of indistinguishable photons (Duarte 1993). In other words, the quantum mechanical description of

N-slit interference for either single-photon propagation or the propagation of indistinguishable photons, *à la Dirac*, is indeed *quantum mechanics*; and, para-phrasing Lamb, it is time we 'learn to enjoy it'.

The concepts introduced by Dirac, Feynman, and Lamb, which are integrated here as the Dirac–Feynman–Lamb (DFL) doctrine, provide the inspiration to apply quantum mechanics to physical processes involving the interaction of populations of indistinguishable photons with macroscopic apparatus. In this regard, the realm of laser interference is particularly apt to the application of quantum physics (Duarte 1993, 2003). The DFL doctrine hints that *this is a quantum world* and encourages the search for new applications of quantum mechanics.

29.6 The importance of the probability amplitude

The concept of the probability amplitude is utterly crucial in quantum mechanics. In this regard, it is the probability amplitude that must faithfully express the physics at hand. From an interferometric perspective, single-photon propagation, and the propagation of a population of indistinguishable photons, an N-slit interferometer is faithfully represented by the Dirac–Feynman probability amplitude

$$\langle x|s \rangle = \sum_{j=1}^{N} \langle x|j \rangle \langle j|s \rangle \tag{29.8}$$

(Duarte 1993). If an extra plane k is introduced, between the N-slit array j and the interferometric plane x, then the probability amplitude immediately becomes

$$\langle x|s \rangle = \sum_{k=1}^{N} \sum_{j=1}^{N} \langle x|k \rangle \langle k|j \rangle \langle j|s \rangle. \tag{29.9}$$

If that extra plane is reduced to a single microscopic and transparent fiber f, then the probability amplitude is immediately reduced to

$$\langle x|s \rangle = \sum_{j=1}^{N} \langle x|f \rangle \langle f|j \rangle \langle j|s \rangle. \tag{29.10}$$

The point here is that any attempt to observe, or to gauge, the original interferogram represented in equation (29.8) immediately results in an entirely new probability amplitude, thus irrevocably altering the original probability amplitude and hence the measured interferogram.

In order to transition from a purely mathematical probability amplitude to a *probability*, then $\langle x|s \rangle$ must be multiplied with its complex conjugate $\langle x|s \rangle^*$ according to Born's rule (Born 1926), that is

$$\langle x|s \rangle \langle x|s \rangle^* = \left(\sum_{j=1}^{N} \langle x|j \rangle \langle j|s \rangle \right) \left(\sum_{j=1}^{N} \langle x|j \rangle \langle j|s \rangle \right)^*. \tag{29.11}$$

Representation of $\langle j|s \rangle$ and $\langle x|j \rangle$ with appropriate complex wave functions, as explained in chapter 2, leads to

$$|\langle x|s \rangle|^2 = \sum_{j=1}^{N} \Psi(r_j) \sum_{m=1}^{N} \Psi(r_m) e^{i(\Omega_m - \Omega_j)} \tag{29.12}$$

and ultimately to (Duarte 1993)

$$|\langle x|s \rangle|^2 = \sum_{j=1}^{N} \Psi(r_j)^2 + 2\sum_{j=1}^{N} \Psi(r_j) \left(\sum_{m=j+1}^{N} \Psi(r_m) \cos(\Omega_m - \Omega_j) \right). \tag{29.13}$$

In this regard, it should be emphasized that equations (29.11)–(29.13) are completely equivalent. Furthermore, equation (29.11) is utterly quantum and has no equivalent whatsoever in classical optics.

Equations (29.11)–(29.13) accurately describe N-slit interference for single-photon propagation and indistinguishable-photon propagation. Here the single photon, or the population of indistinguishable photons, refer to a single-frequency ν, or single-wavelength λ, that in practice can be obtained from a very narrow-linewidth, or a highly-monochromatic, laser. That is, lasers that exhibit a very narrow $\Delta \nu$ or $\Delta \lambda$. In this regard, equation (29.13) should, in more specific terms, be re-expressed as

$$|\langle x|s \rangle|_{\lambda}^2 = \sum_{j=1}^{N} \Psi(r_j)_{\lambda}^2 + 2\sum_{j=1}^{N} \Psi(r_j)_{\lambda} \left(\sum_{m=j+1}^{N} \Psi(r_m)_{\lambda} \cos(\Omega_m - \Omega_j) \right). \tag{29.14}$$

For broadband or semicoherent radiation involving a large manifold of wavelengths $\lambda_1, \lambda_2, \lambda_3, \ldots \lambda_n$, the overall probability becomes (Duarte 2015)

$$\sum_{\lambda=\lambda_1}^{\lambda_n} |\langle x|s \rangle|_{\lambda}^2 = \sum_{\lambda=\lambda_1}^{\lambda_n} \left(\sum_{j=1}^{N} \Psi(r_j)_{\lambda}^2 + 2\sum_{j=1}^{N} \Psi(r_j)_{\lambda} \left(\sum_{m=j+1}^{N} \Psi(r_m)_{\lambda} \cos(\Omega_m - \Omega_j) \right) \right). \tag{29.15}$$

For semicoherent radiation, it is this probability distribution that determines the measured intensity distribution, as outlined in chapter 2 (Duarte 2004). In this regard the spatial detector (either a photographic plate, CCD array, or CMOS spatial detector), records an interferometric distribution corresponding to each wavelength in the $\lambda_1, \lambda_2, \lambda_3, \ldots \lambda_n$ array and provides a *cumulative* intensity distribution. Ample examples of excellent agreement between these equations and measured interferograms, for N in the $2 \leqslant N \leqslant 2000$ range and for ample propagation distances, from the near to the far field, are given in the literature (Duarte 1993, 2015).

The message here is that quantum mechanics not only beautifully, efficiently, and accurately describes macroscopic interference phenomena but also describes the single-photon interference that is completely beyond the scope of classical descriptions of interference. This is one transparent example of the DFL doctrine at work. A comparison between classical and quantum interference is given in appendix C.

Beyond this pragmatic perspective it should be added that equation (29.14) has been applied to accurately predict single-slit diffraction (Duarte 1993, 2015) and to describe interference, diffraction, refraction, and reflection (in that particular hierarchy and order) in a unified, succinct, and rational approach (Duarte 1997, 2015); see appendix I. All of this is described from the Dirac–Feynman probability amplitude!

29.7 The best interpretation of quantum mechanics

Writings on the interpretation of quantum mechanics are rich in philosophical terms, sophisticated words, and uncommon words. In this regard, it should not be forgotten that quantum mechanics was born from the depths of the macroscopic experimental observations of Planck. Observations that led to $E = h\nu$. It should also be remembered that Born emphasized the importance of experimental measurements in the 'tedious' crafting of the laws of quantum mechanics. Also relevant are the words of Dirac: 'I owe a lot to my engineering training because it did teach me to tolerate approximations' (Dalitz 1987). This Dirac acknowledgement should be added to Feynman's observation: 'already in classical mechanics there was indeterminability' (Feynman *et al* 1965). This indeterminacy can be traced all the way back to Newton's *Principia* (Newton 1686).

After contemplating the concepts expressed above and Feynman's thought 'this is the way nature really *is*' (Feynman *et al* 1965), it becomes eventually transparent that the 'least action' approach to interpretational matters, in addition to the laws and principles given in section 29.4, is a *pragmatic* posture devoid of philosophical prejudices. Better still, it can be argued that the most efficient and practical interpretation of quantum mechanics is ... *no interpretation at all* (Duarte 2014).

29.8 Discussion

As described in chapter 17, the fundamental Diracian probability amplitude

$$\langle x|s \rangle = \sum_{j=1}^{N} \langle x|j \rangle \langle j|s \rangle$$

can be utilized as the very foundation to derive in a straightforward and transparent manner the quantum entanglement probability amplitude (Duarte 2013a, 2013b, 2014) first introduced by Pryce and Ward (1947) and Ward (1949):

$$|\psi\rangle = \frac{1}{\sqrt{2}}(|x\rangle_1 \, |y\rangle_2 - |y\rangle_1 \, |x\rangle_2). \tag{29.16}$$

The transition between these two probability amplitudes is quite natural and does not require any mysterious steps or strange concepts. More specifically, it does not invoke any *hidden* theories. Thus it can be stated that, from an interferometric perspective, there are no mysteries in the physics of quantum entanglement.

This perspective is entirely consistent with a pragmatic approach to the interpretation of quantum mechanics and even with the more minimalist principle of invoking no interpretation at all.

References

Andersen A, Madsen J, Reichelt C, Ahl S R, Lautrup B, Ellegaard C, Levinsen M T and Bohr T 2015 Double-slit experiment with single wave-driven particles and its relation to quantum mechanics *Phys. Rev.* E **92** 013006

Bell J S and Nauenberg M 1966 The moral aspects of quantum mechanics *Preludes in Theoretical Physics* ed A De Shalit, H Feshbach and L Van Hove (North Holland: Amsterdam), pp 279–86

Bell J S 1990 Against 'measurement' *Phys. World* **3** 33–40

Bohm D 1952 A suggested interpretation of quantum theory using 'hidden' variables. I *Phys. Rev.* **85** 166–79

Bohm D and Bub J 1966 A proposed solution of the measurement problem in quantum mechanics by hidden variable theory *Rev. Mod. Phys.* **38** 453–69

Bohr N 1935 Can quantum mechanical description of physical reality be considered complete? *Phys. Rev.* **48** 696–702

Born M 1926 Zur quantenmechanik der stoßvorgänge *Z. Phys.* **37** 863–7

Born M 1948 *Natural Philosophy of Cause and Chance* (Oxford: Clarendon)

Dalitz R H 1987 Another side to Paul Dirac *Paul Adrien Maurice Dirac* ed B N Kursunoglu and E P Wigner (Cambridge: Cambridge University Press) ch 10

Dirac P A M 1978 *The Principles of Quantum Mechanics* 4th edn (Oxford: Oxford University Press)

Dirac P A M 1987 The inadequacies of quantum field theory *Paul Adrien Maurice Dirac* ed B N Kursunoglu and E P Wigner (Cambridge: Cambridge University Press) ch 15

Duarte F J 1993 On a generalized interference equation and interferometric measurements *Opt. Commun.* **103** 8–14

Duarte F J 1997 Interference, diffraction, and refraction, via Dirac's notation *Am. J. Phys.* **65** 637–40

Duarte F J 2003 *Tunable Laser Optics* (New York: Elsevier)

Duarte F J 2004 Comment on 'Reflection, refraction and multislit interference' *Eur. J. Phys.* **25** L57–8

Duarte F J 2013a The probability amplitude for entangled polarizations: an interferometric approach *J. Mod. Opt.* **60** 1585–7

Duarte F J 2013b Tunable laser optics: applications to optics and quantum optics *Prog. Quantum Electron.* **37** 326–47

Duarte F J 2014 *Quantum Optics for Engineers* (New York: CRC)

Duarte F J 2015 *Tunable Laser Optics* 2nd edn. (New York: CRC)

Duarte F J, Taylor T S, Clark A B and Davenport W E 2010 The *N*-slit interferometer: an extended configuration *J. Opt.* **12** 015705

Duarte F J, Taylor T S, Black A M, Davenport W E and Vermette P G 2011 N-slit interferometer for secure free-space optical communications: 527 m intra interferometric path length *J. Opt.* **13** 035710

Einstein A, Podolsky B and Rosen N 1935 Can quantum mechanical description of physical reality be considered complete? *Phys. Rev.* **47** 777–80

Feynman R P 1998 *The Meaning of it All* (Reading, MA: Addison-Wesley)

Feynman R P and Hibbs A R 1965 *Quantum Mechanics and Path Integrals* (New York: McGraw-Hill)

Feynman R P, Leighton R B and Sands M 1965 *The Feynman Lectures on Physics* vol III (Reading, MA: Addison-Wesley)

Gottfried K 1991 Does quantum mechanics carry the seeds of its own destruction? *Phys. World* **4** 35–41

Heisenberg W 1927 Über den anschaulichen inhalt der quantentheoretishen kinematik und mechanic *Z. Phys.* **43** 172–98

Kaiser D 2005 *Drawing Theories Apart: The Dispersion of Feynman Diagrams in Post War Physics* (Chicago, IL: The University of Chicago)

Lamb W E 1987 Schrödingers's cat *Paul Adrien Maurice Dirac* ed B N Kursunoglu and E P Wigner (Cambridge: Cambridge University Press) ch 21

Landau L D and Lifshitz E M 1977 *Quantum Mechanics* (New York: Pergamon)

Newton I 1686 *Principia Mathematica* (London: Royal Society)

Planck M 1901 Ueber das gesetz der energieverteilung im normalspectrum *Ann. Phys.* **309** 553–63

Pryce M L H and Ward J C 1947 Angular correlation effects with annihilation radiation *Nature* **160** 435

Schrödinger E 1935 Discussion of probability relations between separated systems *Math. Proc. Camb. Philos. Soc.* **31** 555–63

Schrödinger E 1936 Probability relations between separated systems *Math. Proc. Camb. Philos. Soc.* **32** 446–52

van Kampen N G 1988 Ten theorems about quantum mechanical measurements *Physica* A **153** 97–113

von Neumann J 1932 *Mathematische Grundlagen der Quanten-Mechanik* (Berlin: Springer)

Ward J C 1949 *Some Properties of the Elementary Particles* (Oxford: Oxford University)

Ward J C 2004 *Memoirs of a Theoretical Physicist* (New York: Optics Journal)

Yin J *et al* 2017 Satellite-based entanglement distribution over 1200 kilometers *Science* **356** 1140–4

IOP Publishing

Fundamentals of Quantum Entanglement

F J Duarte

Appendix A

Revisiting the Einstein–Podosky–Rosen (EPR) paper

The EPR argument is examined directly under the premises of Heisenberg's uncertainty principle $\Delta x \Delta p \approx h$.

A.1 Introduction

The contents of the EPR paper entitled 'Can quantum mechanical description of physical reality be considered complete?' (Einstein *et al* 1935) were challenged by Niels Bohr also in a paper entitled 'Can quantum mechanical description of physical reality be considered complete?' (Bohr 1935). In that six-page paper, Bohr centered his argument on Heisenberg's uncertainty principle (Heisenberg 1927)

$$\Delta x \Delta p \approx h \tag{A.1}$$

and on the *complementarity principle*. Bohr's argument, although extensive, apparently failed to convince sectors within the physics community that continued to doubt the completeness of quantum mechanics for decades. Nevertheless, Bohr invoked the two crucial words that led to the one equation that can be applied to neutralize the EPR argument: *uncertainty principle*.

In this regard, it should be mentioned that Dirac already in the 1947 edition of his celebrated book included a remark of extraordinary significance: 'it is evident physically that a state for which all values of q are equally probable, or one for which all values of p are equally probable, cannot be attained in practice' (Dirac 1978).

A.2 EPR and the uncertainty principle

There is a key sentence in the first part of the EPR paper: '*when the momentum of a particle is known, its coordinate has no physical reality*' (Einstein *et al* 1935). A direct confrontation of this central concept with the uncertainty principle leads to an interesting result. The explicit argument that follows is based on concepts previously articulated by Duarte (2014).

doi:10.1088/2053-2563/ab2b33ch30

Heisenberg's uncertainty principle stated in its alternative fractional form (Feynman *et al* 1965) is

$$\Delta x \approx \frac{h}{\Delta p}. \qquad (A.2)$$

Measurement of the momentum of a particle p can only be performed according to

$$p \pm \Delta p. \qquad (A.3)$$

An *absolutely exact measurement of momentum* p with $\Delta p = 0$ is *physically impossible* (Duarte 2014). In this regard, it should be mentioned that uncertainties and errors in measurements have been known to exist since the dawn physics (Newton 1686, 1704). The EPR sentence '*when the momentum of a particle is known, its coordinate has no physical reality*' (Einstein *et al* 1935) implies an exact and perfect measurement of momentum p with $\Delta p = 0$, which is a physical impossibility.

A real non-idealized measurement of momentum leads to $p \pm \Delta p$ with a specific and real non-zero Δp. Once Δp is available, then Δx can be found according to

$$\Delta x \approx \frac{h}{\Delta p}.$$

In this regard, the '*all values*' spread in the coordinate, as feared by Einstein *et al* (1935), is *not allowed*. Removal of the 'all values' spread in the coordinate x immediately neutralizes the claim of '*no physical reality*'. Hence, the EPR conclusion that '*the quantum mechanical description of physical reality ... is not complete*' can be dismissed.

A.3 Conclusion

Here, it has been shown that Heisenberg's uncertainty principle can be effectively applied in a direct and transparent manner to counter EPR's 'all values' argument that led those authors to the conclusion that the description of reality as given by wave functions, or probability amplitudes, 'is not complete'. In this regard, 'the uncertainty principle "protects" quantum mechanics' (Feynman *et al* 1965).

The dismissal of the EPR argument, or the *EPR paradox*, as referred to by many authors, has a profound meaning since it was the EPR argument that led to the formulation of hidden variable theories as presented by Bohm and colleagues (Bohm 1952, Bohm and Bub 1966) and the eventual derivation of Bell's theorem (Bell 1964). This theme is given further discussion in chapter 28.

References

Bell J S 1964 On the Einstein–Podolsky–Rosen paradox *Physics* **1** 195–200

Bohm D 1952 A suggested interpretation of quantum theory using 'hidden' variables. I *Phys. Rev.* **85** 166–79

Bohm D and Bub J 1966 A proposed solution of the measurement problem in quantum mechanics by hidden variable theory *Rev. Mod. Phys.* **38** 453–69

Bohr N 1935 Can quantum mechanical description of physical reality be considered complete? *Phys. Rev.* **48** 696–702

Dirac P A M 1978 *The Principles of Quantum Mechanics* 4th edn (Oxford: Oxford University Press)

Duarte F J 2014 *Quantum Optics for Engineers* (New York: CRC)

Einstein A, Podolsky B and Rosen N 1935 Can quantum mechanical description of physical reality be considered complete? *Phys. Rev.* **47** 777–80

Feynman R P, Leighton R B and Sands M 1965 *The Feynman Lectures on Physics* vol III (Reading, MA: Addison-Wesley)

Heisenberg W 1927 Über den anschaulichen inhalt der quantentheoretischen kinematik und mechanic *Z. Phys.* **43** 172–98

Newton I 1686 *Principia Mathematica* (London: Royal Society)

Newton I 1704 *Opticks* (London: Royal Society)

Appendix B

Revisiting the Pryce–Ward probability amplitude

The work of Pryce and Ward leading to the probability amplitude, $|\psi\rangle = (|x\rangle\,|y\rangle - |y\rangle\,|x\rangle)$, is examined from a historical perspective. Ward's parallel interests at the time, on quantum electrodynamics, are brought forward.

B.1 Introduction

Some readers may wonder why neither Maurice Pryce nor John Ward published a separate dedicated journal paper on the probability amplitude for quantum entanglement, $|\psi\rangle = (|x, y\rangle - |y, x\rangle)$, thus leaving Ward's doctoral thesis as the only explicit contemporaneous record of this development. Concurrently, some readers may also wonder why they never championed, or exploited, the ownership of this most crucial equation, as most physicists would do today if confronted by similar circumstances. These are questions that apparently were never asked of John Ward and one aspect of his physics that he never discussed, at least not in the written record. In this chapter, an attempt is made to find an explanation for this apparent ineffable set of affairs. This discussion is based on measured speculation, personal knowledge of the man, and the published record.

B.2 Exciting times and extreme succinctness

According to John Ward, he was introduced to $|\psi\rangle = (|x, y\rangle - |y, x\rangle)$ via discussions with Maurice Pryce. These discussions were generated by the interest on the perpendicularity of the polarization states of two quanta moving in opposite directions (Ward 2004). At the time, John Ward was working on his doctorate under the supervision of Maurice Pryce at Oxford (Pryce and Ward 1947). However, this was not the only focus of Ward's attention at the time. As his doctoral thesis already hinted in its title, 'Some Properties of the Elementary Particles' (Ward 1949), he was also already attracted by particle physics and by quantum electrodynamics in particular. Perhaps the only other human thinking about quantum optics at the time was Dirac himself.

doi:10.1088/2053-2563/ab2b33ch31

An additional preamble forces a departure from physics to introduce some aspects of John Ward's personality that might be relevant: he was a master of succinctness and always got to the point in the most direct possible way. It is as if he were a living demonstration of the *principle of least action*. He was a fairly distant man with few close friends, physicist Richard Dalitz among them. By today's standards he was extremely honest, modest, and hated corruption and the practitioners of corruption. The reader can find further details on Ward in his autobiography (Ward 2004) and in writings about him (Fraser 2008, Close 2011).

The following facts reinforce and add to the concepts already expressed:

1. His paper with Maurice Pryce entitled 'Angular correlation effects with annihilation radiation' (Pryce and Ward 1947) is slightly longer than half a page. Ward says that Pryce initially refused being a co-author and that he only accepted upon Ward's insistence (Ward 2004).

2. His doctoral thesis *Some Properties of the Elementary Particles* (Ward 1949) was a mere 47 pages long and dealt with two subject matters. The first section was entitled 'Polarization effects of annihilation radiation' and the second section was entitled 'Some higher order effects in covariant quantum electrodynamics'.

3. His landmark paper on renormalization theory entitled 'An identity in quantum electrodynamics' (Ward 1950a) was less than half a page long. The importance of this paper to quantum field theory can be summarized via the statements of experienced practitioners in the field: 'the Ward identity ... ensures the *universality of the electromagnetic interaction*' (Greiner and Reinhardt 2009) and 'the proof that QED can be renormalized relied on Ward's Identities ... Ward's Identities lie at the very foundations of renormalization theory' (Close 2011).

4. Another of his papers on renormalization theory entitled 'A convergent non-linear field theory' (Ward 1950b) was about a third of a page long.

   The evidence above suggests the following as factors that may have prevented publication of a sole and dedicated journal disclosure on $|\psi\rangle = (|x, y\rangle - |y, x\rangle)$:

   (a) In the paper that he wrote with Maurice Pryce, what mattered at the time was the final scattering result useful to experimentalists interested in testing the pair theory. In this regard, Pryce and Ward most likely considered $|\psi\rangle = (|x, y\rangle - |y, x\rangle)$ only as a necessary *intermediate step* to reach that final result. Adhering to succinctness most likely dissuaded them from disclosure.

   (b) Ward's attitude appears to express increased appreciation toward this discovery by the time he presented his thesis since he wrote, 'it is essential to derive correctly the state vector which properly describes the state of the two quanta, including their relative polarization' (Ward 1949). However, it is quite possible that Ward qualified his thesis a publication as good as any other, and thus considered the subject closed. As a matter of fact he would author and co-author only some 20 papers in his entire career.

(c) A dedicated disclosure on the subject would have to have been a joint paper with Pryce, but their paths began to diverge around 1949.

(d) In reference to items 2–4 above it is also quite obvious that the attention of the young physicist quickly shifted from quantum optics, a futuristic subject almost not existing at the time, to quantum electrodynamics and renormalization theory, which were the focus and attention of the physics community. Indeed, Ward would go on to co-author some of the papers that took center stage in the development of the Standard Model (Salam and Ward 1959, 1961 1964a, 1964b).

(e) In the 1970s when teaching quantum mechanics, via the *Feynman Lectures on Physics* (Feynman *et al* 1965), he would acknowledge with a shy smile and few words his rendezvous with equations of the form $|\psi\rangle = (|R\rangle - |L\rangle)$. No further details added.

B.3 Conclusion

The matter of $|\psi\rangle = (|x, y\rangle - |y, x\rangle)$ did resurface between this author and Ward in early 2000. His attitude was that everybody knew the score and that eventually it would be recognized as the work of Pryce and Ward.

What did eventually transpire was that those who knew the score were physicists of his generation, such as Richard Dalitz, Willis Lamb, Maurice Pryce, and C-S Wu, but there was a new score being written by a new generation vastly unaware of the origin of $|\psi\rangle = (|x, y\rangle - |y, x\rangle)$.

In this regard, it is fitting to mention a pertinent quote on Ward's physics: ' ... he has drawn attention to fundamental truths, and has laid down basic principles that physicists have followed in subsequent decades, often without knowing it, and generally without quoting him' (Dunhill 1995).

In summary, succinctness, renormalization, the Standard Model, and uncommon honesty were probably factors in preventing a unique and dedicated disclosure of $|\psi\rangle = (|x, y\rangle - |y, x\rangle)$ by John Ward. Championing his own work was not part of his ethos.

References

Close F 2011 *The Infinity Puzzle* (New York: Basic Books)

Dunhill M 1995 Professor John Clive Ward *The Postmaster and The Merton Record* (Oxford: Oxford University), pp 89–90

Feynman R P, Leighton R B and Sands M 1965 *The Feynman Lectures on Physics* vol III (Reading, MA: Addison-Wesley)

Fraser G 2008 *Cosmic Anger* (Oxford: Oxford University Press)

Greiner W and Reinhardt J 2009 *Quantum Electrodynamics* (Berlin: Springer)

Pryce M L H and Ward J C 1947 Angular correlation effects with annihilation radiation *Nature* **160** 435

Salam A and Ward J C 1959 Weak and electromagnetic interactions *Nuovo Cimento* **11** 568–77
Salam A and Ward J C 1961 On a gauge theory of elementary interactions *Nuovo Cimento* **19** 165–70
Salam A and Ward J C 1964a Electromagnetic and weak interactions *Phys. Lett.* **13** 168–71
Salam A and Ward J C 1964b Gauge theory of elementary interactions *Phys. Rev.* **136** B763–8
Ward J C 1949 *Some Properties of the Elementary Particles* (Oxford: Oxford University Press)
Ward J C 1950a An identity in quantum electrodynamics *Phys. Rev.* **78** 182
Ward J C 1950b A convergent non-linear filed theory *Phys. Rev.* **79** 406
Ward J C 2004 *Memoirs of a Theoretical Physicist* (Rochester, New York: Optics Journal)

Appendix C

Classical and quantum interference

The distinctions between classical interference and quantum interference are emphasized.

C.1 Introduction

In this appendix a brief description of classical N-slit interference and quantum N-slit interference is provided. Here it is shown that although interference can be described classically, this description is only an approximation of the subtle experimental interferometric phenomenon.

C.2 The classical interference equation

From Maxwell's electromagnetic theory, Born and Wolf (1999) derived the interference equation for two-slit interference, or Young's interference, as

$$I = I_1 + I_2 + 2(I_1 I_2)^{-1/2} \cos \delta \tag{C.1}$$

which has the same form as the equation given by Michelson (1927) except that Michelson writes it as

$$i = a_1^2 + a_2^2 + 2a_1 a_2 \cos \delta \tag{C.2}$$

where a_1 and a_2 are designated as amplitudes. Using the notation of Born and Wolf (1999), for an array of N-slits the interference equation (C.1) can be extended to

$$I = \sum_{n=1}^{N} I_n + 2 \sum_{n=1}^{N} I_n^{-\frac{1}{2}} \left(\sum_{m=n+1}^{N} I_m^{-\frac{1}{2}} \cos \delta \right) \tag{C.3}$$

where I_1, I_2, $I_3 \ldots I_n$ refer to the *intensities* present at each of the 1, 2, 3…n slits, which are assumed to be uniform and separated by uniform distances. In these equations, δ is the phase angle derived from the interaction of the light illuminating the slits and the geometry of the slits.

The following observations are applicable to these classical equations:

doi:10.1088/2053-2563/ab2b33ch32

1. These are intensity equations.
2. These are not probability equations.
3. These equations are not applicable to single-photon interference.

C.3 The N-slit interferometer

As already discussed in chapters 2 and 26, the N-slit interferometer is perfectly described by the *generalized interferometric probability* in one dimension:

$$\langle x|s\rangle\langle x|s\rangle^* = \left(\sum_{j=1}^{N}\langle x|j\rangle\langle j|s\rangle\right)\left(\sum_{j=1}^{N}\langle x|j\rangle\langle j|s\rangle\right)^* \tag{C.4}$$

$$\langle x|s\rangle\langle x|s\rangle^* = \sum_{j=1}^{N}\Psi(r_j)\sum_{m=1}^{N}\Psi(r_m)e^{i(\Omega_m-\Omega_j)} \tag{C.5}$$

$$\langle x|s\rangle\langle x|s\rangle^* = \sum_{j=1}^{N}\Psi(r_j)^2 + 2\sum_{j=1}^{N}\Psi(r_j)\left(\sum_{m=j+1}^{N}\Psi(r_m)\cos(\Omega_m-\Omega_j)\right). \tag{C.6}$$

These are three equivalent equations that apply to single-photon propagation or to the propagation of ensembles of indistinguishable photons. Equations (C.5) and (C.6) are obtained from (C.4) while using complex wave forms to represent the probability amplitudes (Duarte 1993, 2003) following Dirac's lead (Dirac 1978).

These equations can also be expressed in two and three dimensions as given in chapter 2 (Duarte 1995).

The following observations are applicable to these N-slit quantum probability equations:
1. These are probability equations.
2. These are not intensity equations.
3. These equations describe single-photon interference and interference of populations of indistinguishable photons.

C.4 The difference between classical and quantum interference

In classical interference

$$Maxwell\ equations\ \rightarrow\ \sum_{n=1}^{N}I_n + 2\sum_{n=1}^{N}I_n^{-\frac{1}{2}}\left(\sum_{m=n+1}^{N}I_m^{-\frac{1}{2}}\cos\delta\right). \tag{C.7}$$

In quantum interference

$$\sum_{j=1}^{N}\langle x|j\rangle\langle j|s\rangle\ \rightarrow\ \sum_{j=1}^{N}\Psi(r_j)\sum_{m=1}^{N}\Psi(r_m)e^{i(\Omega_m-\Omega_j)}. \tag{C.8}$$

It should be noticed that using the semi-coherent version for the quantum probability given in equation (C.5), that is equation (2.17), and the definition for intensity given in equation (2.19), the classical equation for interference (equation (C.3)) can be derived. This is one example that illustrates classical physics as an approximation of quantum mechanics.

References

Born M and Wolf E 1999 *Principles of Optics* (Cambridge: Cambridge University Press)

Dirac P A M 1978 *The Principles of Quantum Mechanics* 4th edn (Oxford: Oxford University Press)

Duarte F J 1993 On a generalized interference equation and interferometric measurements *Opt. Comm.* **103** 8–14

Duarte F J 1995 Interferometric imaging *Tunable Laser Applications* ed F J Duarte (New York: Marcel Dekker)

Duarte F J 2003 *Tunable Laser Optics* (New York: Elsevier)

Michelson A A 1927 *Studies in Optics* (Chicago, IL: The University of Chicago)

Appendix D

Interferometers and their probability amplitudes

The probability amplitudes applicable to the Mach–Zehnder interferometer, the Michelson interferometer, the Sagnac interferometer, and the N-slit interferometer are given.

D.1 Introduction

In this appendix the probability amplitudes for the Mach–Zehnder, the Michelson, the Sagnac, and the N-slit interferometer are given via Dirac's notation (Dirac 1978). This treatment follows the notation given by Duarte (2003, 2014). Excellent classical discussions on these interferometers can be found in Michelson (1927) and Steel (1967).

D.2 Interferometers

The Mach–Zehnder, Michelson, and Sagnac interferometers are comprised of beam splitters and mirrors. Here, the beam splitters are assumed to exhibit perfect 50% reflectivity and 50% transmission while the mirrors are assumed to be 100% perfect reflectors.

For a *single beam splitter*, as described in figure D1, the probability amplitude is given by

$$\langle x|s \rangle = \langle x|j' \rangle \langle j'|s \rangle + \langle x|j \rangle \langle j|s \rangle \tag{D.1}$$

where j represents reflection at the beam splitter and j' stands for transmission. This equation ultimately leads to

$$|s \rangle = \frac{1}{\sqrt{2}}(|A \rangle \pm |B \rangle) \tag{D.2}$$

where

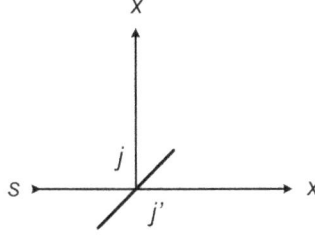

Figure D1. Schematics for a single ideal beam splitter. Interaction of the photon with the beam splitter in the reflection mode is labeled as j while interaction of the photon with the beam splitter in the transmission mode is assigned as j'. Both the transmitted and reflected photons are assumed to be incident on identical detectors x.

$$|A\rangle = |j'\rangle\langle j'|s\rangle \qquad (D.3)$$

and

$$|B\rangle = |j\rangle\langle j|s\rangle \qquad (D.4)$$

provided the two photons are detected by identical detectors x.

D.2.1 The Mach–Zehnder interferometer

Of particular interest to quantum computing is the Mach–Zehnder interferometer, which is comprised of an input beam splitter, an output beam splitter, and two mirrors M_1 and M_2, as illustrated in figure D2. It provides two interferometric outputs, one at x and the other at x'. The interference mechanics of the counter-propagating beams can be described via the following probability amplitude (Duarte 2003, 2014):

$$\langle x|s\rangle = \langle x|k'\rangle\langle k'|M_1\rangle\langle M_1| j\rangle\langle j|s\rangle + \langle x|k\rangle\langle k|M_2\rangle\langle M_2| j'\rangle\langle j'|s\rangle \qquad (D.5)$$

where j and k refer to the beam splitters in the reflective mode while j' and k' refer to the beam splitters in the transmission mode. Assuming perfect reflectivity at mirrors M_1 and M_2, equation (D.5) is equivalent to (Duarte 2003, 2014)

$$\langle x|s\rangle = \langle x|k'\rangle\langle k'| j\rangle\langle j|s\rangle + \langle x|k\rangle\langle k| j'\rangle\langle j'|s\rangle. \qquad (D.6)$$

Using the Dirac identity $|\phi\rangle = |j\rangle\langle j|\phi\rangle$, and abstracting, equation (D.6) reduces to

$$\langle x|s\rangle = \langle x|k'\rangle\langle k'|C\rangle + \langle x|k\rangle\langle k|D\rangle \qquad (D.7)$$

where

$$|D\rangle = |j'\rangle\langle j'|s\rangle \qquad (D.8)$$

and

$$|C\rangle = |j\rangle\langle j|s\rangle. \qquad (D.9)$$

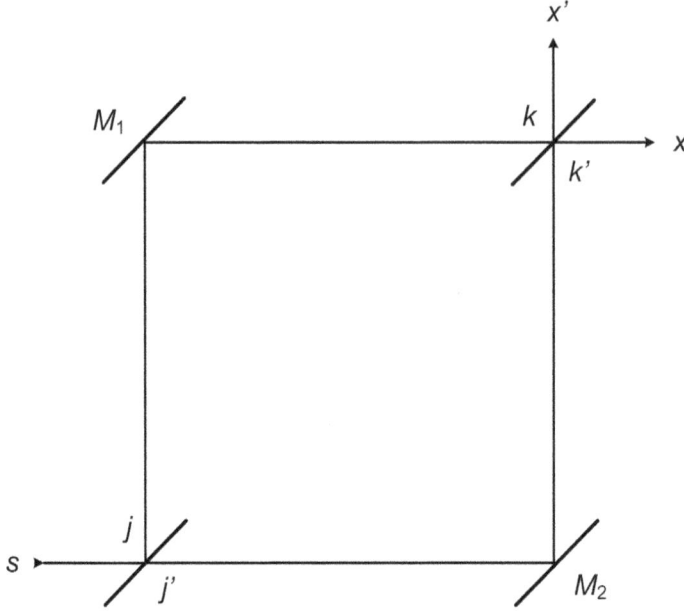

Figure D2. Schematics for the Mach–Zehnder interferometer. Mirrors M_1 and M_2 are assumed to be lossless and perfect. The beam splitters j and k are also assumed to be ideal, lossless, 50–50 partial reflectors. Interaction of the photon with the beam splitters in the reflection mode is labeled as j and k while interaction of the photon with the beam splitter in the transmission mode is assigned as j' and k' (see text).

Then, further abstracting the $\langle x$, and following normalization, equation (D.7) can be reduced to

$$|s\rangle = \frac{1}{\sqrt{2}}(|C\rangle + |D\rangle) \tag{D.10}$$

and once its linear combination is considered the overall probability amplitude becomes

$$|s\rangle = \frac{1}{\sqrt{2}}(|C\rangle \pm |D\rangle). \tag{D.11}$$

For the x' detector,

$$\langle x'|s\rangle = \langle x'|k\rangle\langle k|M_1\rangle\langle M_1|j\rangle\langle j|s\rangle + \langle x'|k'\rangle\langle k'|M_2\rangle\langle M_2|j'\rangle\langle j'|s\rangle \tag{D.12}$$

$$\langle x'|s\rangle = \langle x'|k\rangle\langle k|j\rangle\langle j|s\rangle + \langle x'|k'\rangle\langle k'|j'\rangle\langle j'|s\rangle \tag{D.13}$$

$$\langle x'|s\rangle = \langle x'|k\rangle\langle k|C\rangle + \langle x'|k'\rangle\langle k'|D\rangle \tag{D.14}$$

and ultimately to equation (D.11) again. It should be noticed that if the mirrors M_1 and M_2 are not abstracted, the final result is still given by equation (D.11).

Equation (D.11) gives the probability amplitudes that describe single-photon propagation, or the propagation of an ensemble of indistinguishable photons, in Mach–Zehnder interferometers. It should be noted that given the assumption of perfect mirrors only the beam splitters contribute to the final result. In essence, these equations describe single-photon propagation via two identical beam splitters.

The state $|C\rangle$ *is different* from $|A\rangle$, in equation (D.2), since it includes information about transmission via the first beam splitter, reflection at M_2, and reflection at the second beam splitter. The same observation is valid when comparing $|D\rangle$ to $|B\rangle$.

One final observation is that equation (D.7) can be directly derived from the generalized Dirac probability amplitude

$$\langle x|s \rangle = \sum_{j=1}^{N=2} \langle x|j \rangle \langle j|s \rangle \tag{D.15}$$

for $N = 2$, which is applicable to the double-slit interferometer (see chapter 17). However, physically speaking, a Mach–Zehnder interferometer is very different from a double-slit interferometer. In the double-slit interferometer, the single photon, or the population of indistinguishable photons, undergoes violent diffraction at the slits. This diffraction makes the physics between the two interferometers quite different. The only similarity between the two interferometers is that they are both two-path interferometers, that is, $N = 2$. However, while the Mach–Zehnder interferometer is a *two-beam* interferometer the double-slit, or two-slit, or Young, interferometer is a *parallel diffraction* interferometer.

D.2.2 The Michelson interferometer

The Michelson interferometer (Michelson 1927) is comprised of one beam splitter and two mirrors M_1 and M_2 in an L configuration, as depicted in figure D3. In reference to the schematics, the probability amplitude describing single-photon propagation from the source s to the detector x is given by

$$\langle x|s \rangle = \langle x|j \rangle \langle j|M_2 \rangle \langle M_2|j' \rangle \langle j'|s \rangle + \langle x|j' \rangle \langle j'|M_1 \rangle \langle M_1|j \rangle \langle j|s \rangle \tag{D.16}$$

where j represents reflection at the beam splitter and j' stands for transmission.

Assuming perfect reflectivity, equation (D.16) can be abstracted to

$$\langle x|s \rangle = \langle x|j \rangle \langle j|j' \rangle \langle j'|s \rangle + \langle x|j' \rangle \langle j'|j \rangle \langle j|s \rangle. \tag{D.17}$$

Further abstraction, using $|\phi\rangle = |j\rangle \langle j|\phi\rangle$, leads to

$$\langle x|s \rangle = \langle x|j \rangle \langle j|s \rangle + \langle x|j' \rangle \langle j'|s \rangle \tag{D.18}$$

which again leads to a probability amplitude of the form of equation (D.15).

As seen previously, this equation can be abstracted into

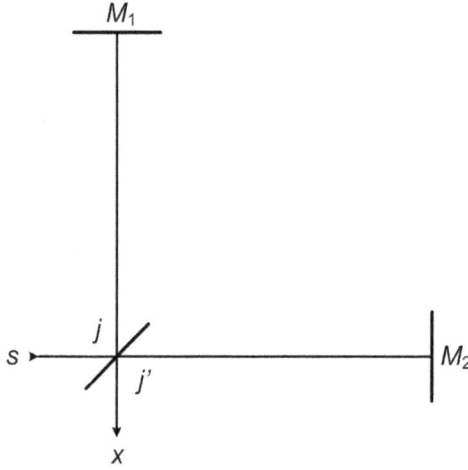

Figure D3. Schematics of the Michelson interferometer.

$$|s\rangle = \frac{1}{\sqrt{2}}(|E\rangle \pm |F\rangle) \tag{D.19}$$

where

$$|F\rangle = |j'\rangle\langle j'|s\rangle \tag{D.20}$$

and

$$|E\rangle = |j\rangle\langle j|s\rangle \tag{D.21}$$

which is not surprising since the mirrors M_1 and M_2 are being treated as idealized perfect mirrors leaving all the physics to the beam splitter. A variant of the Michelson interferometer uses retroreflectors (Steel 1967).

D.2.3 The Sagnac interferometer

The Sagnac interferometer is comprised of one beam splitter and three mirrors, as illustrated in figure D4. Using the same meaning for j and j' as previously, the corresponding probability amplitude can be expressed as

$$\begin{aligned}\langle x|s\rangle &= \langle x|j\rangle\langle j|M_3\rangle\langle M_3|M_2\rangle\langle M_2|M_1\rangle\langle M_1|j\rangle\langle j|s\rangle \\ &+ \langle x|j'\rangle\langle j'|M_1\rangle\langle M_1|M_2\rangle\langle M_2|M_3\rangle\langle M_3|j'\rangle\langle j'|s\rangle.\end{aligned} \tag{D.22}$$

Assuming perfect reflectivity at the mirrors,

$$\langle j|M_3\rangle\langle M_3|M_2\rangle\langle M_2|M_1\rangle\langle M_1|j\rangle = 1 \tag{D.23}$$

$$\langle j'|M_1\rangle\langle M_1|M_2\rangle\langle M_2|M_3\rangle\langle M_3|j'\rangle = 1 \tag{D.24}$$

and equation (D.22) reduces to

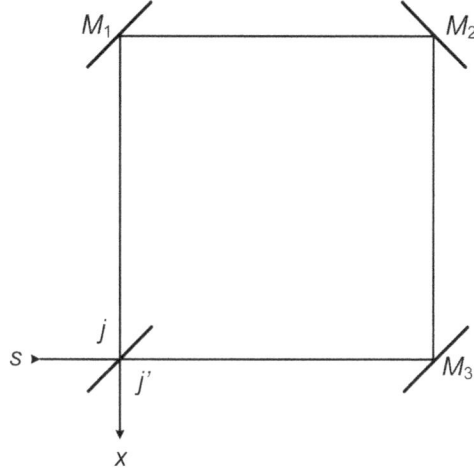

Figure D4. Schematics of the Sagnac interferometer with three mirrors M_1, M_2, and M_3.

$$\langle x|s\rangle = \langle x|j\rangle\langle j|s\rangle + \langle x|j'\rangle\langle j'|s\rangle \tag{D.25}$$

which can ultimately be expressed as

$$|s\rangle = \frac{1}{\sqrt{2}}(|G\rangle \pm |H\rangle) \tag{D.26}$$

where

$$|H\rangle = |j'\rangle\langle j'|s\rangle \tag{D.27}$$

and

$$|G\rangle = |j\rangle\langle j|s\rangle. \tag{D.28}$$

Again, this is due to simplifying assumptions made in equations (D.23) and (D.24).

Furthermore, it should be noted that the physics of equation (D.25) can be traced back to the probability amplitude given in equation (D.15).

The alternative triangular Sagnac interferometer, with only two mirrors (M_1 and M_2), illustrated in figure D5, leads to

$$\langle x|s\rangle = \langle x|j\rangle\langle j|M_2\rangle\langle M_2|M_1\rangle\langle M_1|j\rangle\langle j|s\rangle \\ + \langle x|j'\rangle\langle j'|M_1\rangle\langle M_1|M_2\rangle\langle M_2|j'\rangle\langle j'|s\rangle \tag{D.29}$$

with the same conclusions as with the Sagnac interferometer with three mirrors.

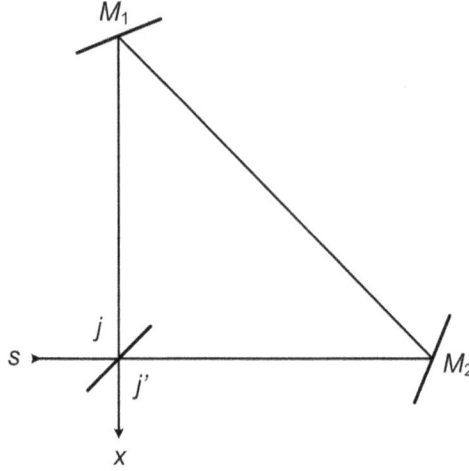

Figure D5. Schematics of the triangular Sagnac interferometer incorporating mirrors M_1 and M_2.

D.2.4 The N-slit interferometer

As already discussed in chapters 2 and 26, the N-slit interferometer is perfectly described by the Dirac–Feynman probability amplitude

$$\langle x|s \rangle = \sum_{j=1}^{N} \langle x|j \rangle \langle j|s \rangle \tag{D.30}$$

and leads the generalized probability in one dimension:

$$\langle x|s \rangle \langle x|s \rangle^* = \left(\sum_{j=1}^{N} \langle x|j \rangle \langle j|s \rangle \right) \left(\sum_{j=1}^{N} \langle x|j \rangle \langle j|s \rangle \right)^* \tag{D.31}$$

$$\langle x|s \rangle \langle x|s \rangle^* = \sum_{j=1}^{N} \Psi(r_j) \sum_{m=1}^{N} \Psi(r_m) e^{i(\Omega_m - \Omega_j)} \tag{D.32}$$

$$\langle x|s \rangle \langle x|s \rangle^* = \sum_{j=1}^{N} \Psi(r_j)^2 + 2 \sum_{j=1}^{N} \Psi(r_j) \left(\sum_{m=j+1}^{N} \Psi(r_m) \cos(\Omega_m - \Omega_j) \right). \tag{D.33}$$

These three equivalent equations apply to single-photon propagation or to the propagation of ensembles of indistinguishable photons. For explicit long-hand versions of equation (D.33) for $N = 2, 3 \ldots 5$, the reader should refer to Duarte (2014, 2015).

For instance, for $N = 7$(Duarte 2015),

$$
\begin{aligned}
|\langle x|s\rangle|^2 = {}& \Psi(r_1)^2 + \Psi(r_2)^2 + \Psi(r_3)^2 + \Psi(r_4)^2 + \Psi(r_5)^2 + \Psi(r_6)^2 + \Psi(r_7)^2 \\
& + 2(\Psi(r_1)\Psi(r_2)\cos(\Omega_2 - \Omega_1) + \Psi(r_1)\Psi(r_3)\cos(\Omega_3 - \Omega_1) \\
& + \Psi(r_1)\Psi(r_4)\cos(\Omega_4 - \Omega_1) + \Psi(r_1)\Psi(r_5)\cos(\Omega_5 - \Omega_1) \\
& + \Psi(r_1)\Psi(r_6)\cos(\Omega_6 - \Omega_1) + \Psi(r_1)\Psi(r_7)\cos(\Omega_7 - \Omega_1) \\
& + \Psi(r_2)\Psi(r_3)\cos(\Omega_3 - \Omega_2) + \Psi(r_2)\Psi(r_4)\cos(\Omega_4 - \Omega_2) \\
& + \Psi(r_2)\Psi(r_5)\cos(\Omega_5 - \Omega_2) + \Psi(r_2)\Psi(r_6)\cos(\Omega_6 - \Omega_2) \\
& + \Psi(r_2)\Psi(r_7)\cos(\Omega_7 - \Omega_2) + \Psi(r_3)\Psi(r_4)\cos(\Omega_4 - \Omega_3) \\
& + \Psi(r_3)\Psi(r_5)\cos(\Omega_5 - \Omega_3) + \Psi(r_3)\Psi(r_6)\cos(\Omega_6 - \Omega_3) \\
& + \Psi(r_3)\Psi(r_7)\cos(\Omega_7 - \Omega_3) + \Psi(r_4)\Psi(r_5)\cos(\Omega_5 - \Omega_4) \\
& + \Psi(r_4)\Psi(r_6)\cos(\Omega_6 - \Omega_4) + \Psi(r_4)\Psi(r_7)\cos(\Omega_7 - \Omega_4) \\
& + \Psi(r_5)\Psi(r_6)\cos(\Omega_6 - \Omega_5) + \Psi(r_5)\Psi(r_7)\cos(\Omega_7 - \Omega_5) \\
& + \Psi(r_6)\Psi(r_7)\cos(\Omega_7 - \Omega_6)).
\end{aligned}
\tag{D.34}
$$

D.3 Beam splitter matrices

A straightforward non-polarizing beam-splitter is a partial mirror. The 2×2 transfer matrix describing the action of a partial reflector, or beam splitter, is simply the identity matrix (Siegman 1986; Duarte 2003)

$$
I = \begin{pmatrix} 1 & 0 \\ 0 & 1 \end{pmatrix}
\tag{D.35}
$$

If a photon polarized in the $|x\rangle$ state encounters a non-polarizing beam splitter, deployed at $\theta = \pi/4$, then there is a probability amplitude for straight passage and a probability amplitude for reflection onto a path orthogonal to initial direction of propagation. *No change in polarization is experienced.* The situation is different when dealing with the *Hadamard matrix* as a beam splitter. The 2×2 Hadamard matrix, which is described as a time symmetric beam splitter, can be expressed as (see chapter 24)

$$
H = 2^{-1/2} \begin{pmatrix} 1 & 1 \\ 1 & -1 \end{pmatrix}
\tag{D.36}
$$

which is equivalent to (see appendix F)

$$
H = (|\psi\rangle_+ + |\psi\rangle^-)
\tag{D.37}
$$

The Hadamard H operating on the $|x\rangle$ and $|y\rangle$ states yields

$$H|x\rangle = 2^{-1/2}(|x\rangle + |y\rangle) \qquad (\text{D.38})$$

$$H|y\rangle = 2^{-1/2}(|x\rangle - |y\rangle) \qquad (\text{D.39})$$

It is immediately clear that equations (D.38) and (D.39) have the same form of equation (D.2).

References

Dirac P A M 1978 *The Principles of Quantum Mechanics* 4th edn (Oxford: Oxford University Press)

Duarte F J 2003 *Tunable Laser Optics* (New York: Elsevier)

Duarte F J 2014 *Quantum Optics for Engineers* (New York: CRC)

Duarte F J 2015 *Tunable Laser Optics* 2nd edn (New York: CRC)

Michelson A A 1927 *Studies in Optics* (Chicago, IL: The University of Chicago)

Siegman A E 1986 *Lasers* (Mill Valley, CA: University Science Books)

Steel W H 1967 *Interferometry* (Cambridge: Cambridge University Press)

Zeilinger A, Bernstein H J and Horne M A 1994 Information transfer with two-state two-particle quantum systems *J. Mod. Opt.* **41** 2375–84

IOP Publishing

Fundamentals of Quantum Entanglement

F J Duarte

Appendix E

Polarization rotators

This is a brief introduction to the matrices used to describe polarization rotators such as wave plates, rhomboids, and broadband prismatic rotators.

E.1 Introduction

Polarization rotation devices are widely utilized in quantum optics. Here, an ultra-brief introduction is given with attention to the matrices governing the rotation. For an excellent theoretical review, the book of Robson (1974) is recommended along with Born and Wolf (1999). A more experimental perspective is given by Duarte (2014).

E.2 Wave plates

The generalized matrix for birefringent rotators is given by (Robson 1974):

$$\begin{pmatrix} \cos\theta & \sin\theta \\ -e^{i\delta}\sin\theta & e^{i\delta}\cos\theta \end{pmatrix}. \tag{E.1}$$

For a quarter-wave plate $\delta = \pi/2$, the phase term is $e^{i\pi/2} = +i$, so that

$$R_{1/4} = \begin{pmatrix} \cos\theta & \sin\theta \\ -i\sin\theta & i\cos\theta \end{pmatrix}. \tag{E.2}$$

For a half-wave plate $\delta = \pi$ and $e^{i\pi} = -1$, so that

$$R_{1/2} = \begin{pmatrix} \cos\theta & \sin\theta \\ \sin\theta & -\cos\theta \end{pmatrix}. \tag{E.3}$$

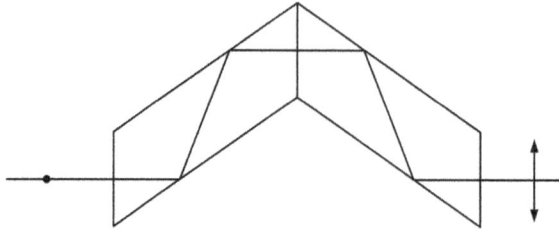

Figure E1. Schematics for a generic double Fresnel rhomb utilized for $\theta = \pi/2$ rotation of linearly polarized light.

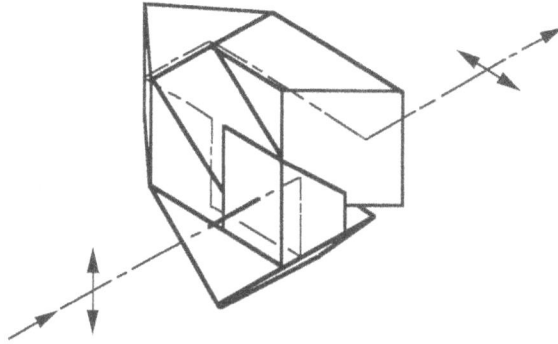

Figure E2. Schematics of the broadband collinear multiple-prism rotator utilized for $\theta = \pi/2$ rotation of linearly polarized light (from Duarte 1989).

Half-wave plates cause rotation of linearly polarized beams by $\theta = \pi/2$ so that the rotation matrix reduces to

$$R_{1/2} = \begin{pmatrix} 0 & 1 \\ 1 & 0 \end{pmatrix}. \tag{E.4}$$

Quarter-wave plates and half-wave plates are wavelength specific. However, their wavelength performance can be improved using achromatic designs incorporating multiple crystal materials.

E.3 Rhomboid and prismatic rotators

Other useful polarization rotators include the wavelength-specific double Fresnel rhomb, illustrated in figure E1, and the broadband collinear multiple-prism rotator (Duarte 1989), displayed in figure E2. Both these rotators turn linearly polarized light by $\theta = \pi/2$ so that their polarization matrix is

$$R = \begin{pmatrix} 0 & 1 \\ 1 & 0 \end{pmatrix}.$$

The broadband collinear multiple-prism rotator has demonstrated high-fidelity transmission for $\theta = \pi/2$ rotation, at efficiencies approaching 95% (Duarte 1992).

References

Duarte F J 1989 Optical device for rotating the polarization of a light beam *US Patent* 4822150

Duarte F J 1992 Beam transmission characteristics of a collinear polarization rotator *Appl. Opt.* **31** 3377–8

Duarte F J 2014 *Quantum Optics for Engineers* (New York: CRC)

Robson B A 1974 *The Theory of Polarization Phenomena* (Oxford: Clarendon)

Born M and Wolf E 1999 *Principles of Optics* (Cambridge: Cambridge University Press)

IOP Publishing

Fundamentals of Quantum Entanglement

F J Duarte

Appendix F

Vector products in quantum notation

Vector products are reviewed. Of particular interest are the vector products relevant to quantum probability amplitudes. These include the product utilized in density matrix calculations, the vector direct product, the vector outer product, and the Kronecker or tensor product. The equivalence in vector notation that represents $|x\rangle$ and $|y\rangle$ polarization states is also illustrated.

F.1 Introduction

In this appendix some aspects of vector algebra, vector products in particular, are described from a direct utilitarian perspective.

For a vector in three dimensions (x, y, z), the sum of two vectors $\mathbf{u} + \mathbf{w}$ is defined as

$$\begin{pmatrix} u_1 \\ u_2 \\ u_3 \end{pmatrix} + \begin{pmatrix} w_1 \\ w_2 \\ w_3 \end{pmatrix} = \begin{pmatrix} u_1 + w_1 \\ u_2 + w_2 \\ u_3 + w_3 \end{pmatrix} \tag{F.1}$$

while subtraction $\mathbf{u} - \mathbf{w}$ is defined as

$$\begin{pmatrix} u_1 \\ u_2 \\ u_3 \end{pmatrix} - \begin{pmatrix} w_1 \\ w_2 \\ w_3 \end{pmatrix} = \begin{pmatrix} u_1 - w_1 \\ u_2 - w_2 \\ u_3 - w_3 \end{pmatrix}. \tag{F.2}$$

Multiplication of a vector $\mathbf{u}$ with a scalar number a, yielding a new vector $a\mathbf{u}$, is defined as

$$a \begin{pmatrix} u_1 \\ u_2 \\ u_3 \end{pmatrix} = \begin{pmatrix} au_1 \\ au_2 \\ au_3 \end{pmatrix}. \tag{F.3}$$

The length of a vector $\mathbf{u}$ is defined as $|\mathbf{u}|$:

$$|\mathbf{u}|^2 = \begin{vmatrix} u_1 \\ u_2 \\ u_3 \end{vmatrix}^2 = \left(u_1^2 + u_2^2 + u_3^2 \right). \tag{F.4}$$

F.2 Vector products

Various vector products are useful in quantum optics. The vector quantum notation is used for the density matrix, the vector direct product, the tensor outer product, and the Kronecker product. The quantum vectors $|u\rangle$ and $|w\rangle$ are two-dimensional and thus compatible with 2×2 matrices.

F.2.1 Dot product

The dot product of two vectors $\mathbf{u} \cdot \mathbf{w}$ is a scalar defined as

$$\begin{pmatrix} u_1 \\ u_2 \\ u_3 \end{pmatrix} \cdot \begin{pmatrix} w_1 \\ w_2 \\ w_3 \end{pmatrix} = (u_1 w_1 + u_2 w_2 + u_3 w_3). \tag{F.5}$$

If the angle between the two vectors is defined as θ,

$$\mathbf{u} \cdot \mathbf{w} = |\mathbf{u}||\mathbf{w}| \cos \theta. \tag{F.6}$$

F.2.2 Cross product

The cross product of two vectors leading to a new vector $\mathbf{u} \times \mathbf{w}$ is defined as

$$\begin{pmatrix} u_1 \\ u_2 \\ u_3 \end{pmatrix} \times \begin{pmatrix} w_1 \\ w_2 \\ w_3 \end{pmatrix} = \begin{pmatrix} u_2 w_3 - u_3 w_2 \\ u_3 w_1 - u_1 w_3 \\ u_1 w_2 - u_2 w_1 \end{pmatrix}. \tag{F.7}$$

F.2.3 Density matrix

The density matrix is defined as the product of two vectors, a *bra* vector and a *ket* vector (Dirac 1978):

$$\rho = |u\rangle \langle u| \tag{F.8}$$

$$\rho = |u\rangle \langle u| = \begin{pmatrix} u_1 \\ u_2 \end{pmatrix} (u_1^* \ u_2^*) = \begin{pmatrix} u_1 u_1^* & u_1 u_2^* \\ u_2 u_1^* & u_2 u_2^* \end{pmatrix}. \tag{F.9}$$

F.2.4 Vector direct product

Notice that this is different from the *dot product* (see, for example, Ayres 1965):

$$|u\rangle |w\rangle = |u\rangle \cdot |w\rangle^T \tag{F.10}$$

$$|u\rangle \cdot |w\rangle^T = \begin{pmatrix} u_1 \\ u_2 \end{pmatrix} (w_1 \quad w_2) = \begin{pmatrix} u_1 w_1 & u_1 w_2 \\ u_2 w_1 & u_2 w_2 \end{pmatrix}. \tag{F.11}$$

F.2.5 Vector outer product

The vector outer product is sometimes associated with the symbol $\otimes$. However, it is handled mechanically as the direct product (Ortega 1987)

$$|u\rangle \, |w\rangle^T = |u\rangle \otimes |w\rangle \tag{F.12}$$

$$|u\rangle \otimes |w\rangle = \begin{pmatrix} u_1 \\ u_2 \end{pmatrix} (w_1 \quad w_2) = \begin{pmatrix} u_1 w_1 & u_1 w_2 \\ u_2 w_1 & u_2 w_2 \end{pmatrix}. \tag{F.13}$$

The symbol $\otimes$ is also used for the Kronecker product, leading sometimes to confusion.

F.2.6 Kronecker product or tensor product

The Kronecker product, $\mathbf{U} \otimes \mathbf{W}$, is a form of matrix multiplication in which each element of the product matrix is comprised of each of the elements of the $\mathbf{U}$ matrix, u_{mn}, multiplying the whole $\mathbf{W}$ matrix, so that the first element is $u_{11}\,\mathbf{W}$, and the last element is $u_{mn}\,\mathbf{W}$ (Zehfuss 1858). This means, for example, that the Kronecker product of two 2×2 matrices yields a 4×4 matrix. For simple two-dimensional vectors this product can be expressed as

$$|u\rangle \otimes |w\rangle = \begin{pmatrix} u_1 \\ u_2 \end{pmatrix} \otimes \begin{pmatrix} w_1 \\ w_2 \end{pmatrix} = \begin{pmatrix} u_1 w_1 \\ u_1 w_2 \\ u_2 w_1 \\ u_2 w_2 \end{pmatrix}. \tag{F.14}$$

For

$$|1\rangle = \begin{pmatrix} 1 \\ 0 \end{pmatrix} \tag{F.15}$$

$$|0\rangle = \begin{pmatrix} 0 \\ 1 \end{pmatrix} \tag{F.16}$$

the following Kronecker products follow:

$$|1\rangle \otimes |1\rangle = \begin{pmatrix} 1 \\ 0 \\ 0 \\ 0 \end{pmatrix} \tag{F.17}$$

$$|1\rangle \otimes |0\rangle = \begin{pmatrix} 0 \\ 1 \\ 0 \\ 0 \end{pmatrix} \tag{F.18}$$

$$|0\rangle \otimes |1\rangle = \begin{pmatrix} 0 \\ 0 \\ 1 \\ 0 \end{pmatrix} \tag{F.19}$$

$$|0\rangle \otimes |0\rangle = \begin{pmatrix} 0 \\ 0 \\ 0 \\ 1 \end{pmatrix}. \tag{F.20}$$

Using the Kronecker product on the Pauli matrices yields the following 4×4 matrices:

$$\sigma_x \otimes \sigma_y = \begin{pmatrix} 0 & 1 \\ 1 & 0 \end{pmatrix} \otimes \begin{pmatrix} 0 & -i \\ i & 0 \end{pmatrix} = \begin{pmatrix} 0 & 0 & 0 & -i \\ 0 & 0 & i & 0 \\ 0 & -i & 0 & 0 \\ i & 0 & 0 & 0 \end{pmatrix} \tag{F.21}$$

$$\sigma_x \otimes \sigma_z = \begin{pmatrix} 0 & 1 \\ 1 & 0 \end{pmatrix} \otimes \begin{pmatrix} 1 & 0 \\ 0 & -1 \end{pmatrix} = \begin{pmatrix} 0 & 0 & 1 & 0 \\ 0 & 0 & 0 & -1 \\ 1 & 0 & 0 & 0 \\ 0 & -1 & 0 & 0 \end{pmatrix} \tag{F.22}$$

$$\sigma_y \otimes \sigma_x = \begin{pmatrix} 0 & -i \\ i & 0 \end{pmatrix} \otimes \begin{pmatrix} 0 & 1 \\ 1 & 0 \end{pmatrix} = \begin{pmatrix} 0 & 0 & 0 & -i \\ 0 & 0 & -i & 0 \\ 0 & i & 0 & 0 \\ i & 0 & 0 & 0 \end{pmatrix} \tag{F.23}$$

$$\sigma_y \otimes \sigma_z = \begin{pmatrix} 0 & -i \\ i & 0 \end{pmatrix} \otimes \begin{pmatrix} 1 & 0 \\ 0 & -1 \end{pmatrix} = \begin{pmatrix} 0 & 0 & -i & 0 \\ 0 & 0 & 0 & i \\ i & 0 & 0 & 0 \\ 0 & -i & 0 & 0 \end{pmatrix} \tag{F.24}$$

$$\sigma_z \otimes \sigma_x = \begin{pmatrix} 1 & 0 \\ 0 & -1 \end{pmatrix} \otimes \begin{pmatrix} 0 & 1 \\ 1 & 0 \end{pmatrix} = \begin{pmatrix} 0 & 1 & 0 & 0 \\ 1 & 0 & 0 & 0 \\ 0 & 0 & 0 & -1 \\ 0 & 0 & -1 & 0 \end{pmatrix} \tag{F.25}$$

$$\sigma_z \otimes \sigma_y = \begin{pmatrix} 1 & 0 \\ 0 & -1 \end{pmatrix} \otimes \begin{pmatrix} 0 & -i \\ i & 0 \end{pmatrix} = \begin{pmatrix} 0 & -i & 0 & 0 \\ i & 0 & 0 & 0 \\ 0 & 0 & 0 & i \\ 0 & 0 & -i & 0 \end{pmatrix}. \tag{F.26}$$

F.3 Equivalence in vector notation for entangled polarizations

Here, for the sake of transparency, a clarification in the definition of vector notation is made explicit. In the notation utilized in this monograph, the polarization $|x\rangle$ and $|y\rangle$ states are represented by $|1\rangle$ and $|0\rangle$, and their corresponding vectors, as defined by Fowles (1968) and Robson (1974), are

$$|x\rangle = |1\rangle = \begin{pmatrix} 1 \\ 0 \end{pmatrix} \tag{F.27}$$

and

$$|y\rangle = |0\rangle = \begin{pmatrix} 0 \\ 1 \end{pmatrix}. \tag{F.28}$$

However, in the contemporaneous literature (see, for example, Nielsen and Chuang 2000) the convention

$$|x\rangle = |0\rangle = \begin{pmatrix} 1 \\ 0 \end{pmatrix} \tag{F.29}$$

and

$$|y\rangle = |1\rangle = \begin{pmatrix} 0 \\ 1 \end{pmatrix} \tag{F.30}$$

is used. The point to be made here is that both conventions are equivalent as long as consistency is maintained.

To illustrate the validity of the previous statement, first the definition expressed in equations (F.27) and (F. 28) is used in

$$|\psi\rangle_+ = 2^{-1/2}(|1\rangle\,|0\rangle + |0\rangle\,|1\rangle) \tag{F.31}$$

$$|\psi\rangle_- = 2^{-1/2}(|1\rangle\,|0\rangle - |0\rangle\,|1\rangle) \tag{F.32}$$

$$|\psi\rangle^+ = 2^{-1/2}(|1\rangle\,|1\rangle + |0\rangle\,|0\rangle) \tag{F.33}$$

$$|\psi\rangle^- = 2^{-1/2}(|1\rangle\,|1\rangle - |0\rangle\,|0\rangle) \tag{F.34}$$

to yield, using the vector direct product,

$$|\psi\rangle_+ = 2^{-1/2}\left(\begin{pmatrix}1\\0\end{pmatrix}\cdot\begin{pmatrix}0\\1\end{pmatrix} + \begin{pmatrix}0\\1\end{pmatrix}\cdot\begin{pmatrix}1\\0\end{pmatrix}\right) = 2^{-1/2}\left(\begin{pmatrix}0&1\\0&0\end{pmatrix} + \begin{pmatrix}0&0\\1&0\end{pmatrix}\right)$$
$$= 2^{-1/2}\begin{pmatrix}0&1\\1&0\end{pmatrix} \tag{F.35}$$

$$|\psi\rangle_- = 2^{-1/2}\left(\begin{pmatrix}1\\0\end{pmatrix}\cdot\begin{pmatrix}0\\1\end{pmatrix} - \begin{pmatrix}0\\1\end{pmatrix}\cdot\begin{pmatrix}1\\0\end{pmatrix}\right) = 2^{-1/2}\left(\begin{pmatrix}0&1\\0&0\end{pmatrix} - \begin{pmatrix}0&0\\1&0\end{pmatrix}\right)$$
$$= 2^{-1/2}\begin{pmatrix}0&1\\-1&0\end{pmatrix} \tag{F.36}$$

$$|\psi\rangle^+ = 2^{-1/2}\left(\begin{pmatrix}1\\0\end{pmatrix}\cdot\begin{pmatrix}1\\0\end{pmatrix} + \begin{pmatrix}0\\1\end{pmatrix}\cdot\begin{pmatrix}0\\1\end{pmatrix}\right) = 2^{-1/2}\left(\begin{pmatrix}1&0\\0&0\end{pmatrix} + \begin{pmatrix}0&0\\0&1\end{pmatrix}\right)$$
$$= 2^{-1/2}\begin{pmatrix}1&0\\0&1\end{pmatrix} \tag{F.37}$$

$$|\psi\rangle^- = 2^{-1/2}\left(\begin{pmatrix}1\\0\end{pmatrix}\cdot\begin{pmatrix}1\\0\end{pmatrix} - \begin{pmatrix}0\\1\end{pmatrix}\cdot\begin{pmatrix}0\\1\end{pmatrix}\right) = 2^{-1/2}\left(\begin{pmatrix}1&0\\0&0\end{pmatrix} - \begin{pmatrix}0&0\\0&1\end{pmatrix}\right)$$
$$= 2^{-1/2}\begin{pmatrix}1&0\\0&-1\end{pmatrix} \tag{F.38}$$

which can be summarized in the following identities (Duarte *et al* 2019)

$$|\psi\rangle_+ = 2^{-1/2}\sigma_x \tag{F.39}$$

$$|\psi\rangle_- = 2^{-1/2}i\sigma_y \tag{F.40}$$

$$|\psi\rangle^+ = 2^{-1/2}I \tag{F.41}$$

$$|\psi\rangle^- = 2^{-1/2}\sigma_z. \tag{F.42}$$

Now, using instead the definitions of equations (F.29) and (F.30) and

$$|\psi\rangle_+ = 2^{-1/2}(|0\rangle\,|1\rangle + |1\rangle\,|0\rangle) \tag{F.43}$$

$$|\psi\rangle_- = 2^{-1/2}(|0\rangle\,|1\rangle - |1\rangle\,|0\rangle) \tag{F.44}$$

$$|\psi\rangle^+ = 2^{-1/2}(|0\rangle\,|0\rangle + |1\rangle\,|1\rangle) \tag{F.45}$$

$$|\psi\rangle^- = 2^{-1/2}(|0\rangle\,|0\rangle - |1\rangle\,|1\rangle) \tag{F.46}$$

the reader can verify that equations (F.35)–(F.38) are again reproduced, and so are the identities expressed in (F.39)–(F.42).

Furthermore, using the definition of equations (F.27) and (F.28)

$$\sigma_z |1\rangle = |1\rangle \tag{F.47}$$

$$\sigma_z |0\rangle = -|0\rangle \tag{F.48}$$

but, using the definitions of equations (F.29) and (F.30)

$$\sigma_z |0\rangle = |0\rangle \tag{F.49}$$

$$\sigma_z |1\rangle = -|1\rangle. \tag{F.50}$$

However, the explicit versions of equations (F.47) and (F.48) are

$$\begin{pmatrix} 1 & 0 \\ 0 & -1 \end{pmatrix}\begin{pmatrix} 1 \\ 0 \end{pmatrix} = \begin{pmatrix} 1 \\ 0 \end{pmatrix} \tag{F.51}$$

$$\begin{pmatrix} 1 & 0 \\ 0 & -1 \end{pmatrix}\begin{pmatrix} 0 \\ 1 \end{pmatrix} = -\begin{pmatrix} 0 \\ 1 \end{pmatrix} \tag{F.52}$$

which, as the reader can verify, are the same as the explicit versions of equations (F.49) and (F.50).

F.4 The Hadamard matrix and quantum entanglement

The nexus between the Hadamard matrix and the probabilities for quantum entanglement can be elucidated by considering equations (F.35) and (F.38) in their final form

$$|\psi\rangle_+ = 2^{-1/2}\begin{pmatrix} 0 & 1 \\ 1 & 0 \end{pmatrix} \tag{F.53}$$

$$|\psi\rangle^- = 2^{-1/2}\begin{pmatrix} 1 & 0 \\ 0 & -1 \end{pmatrix} \tag{F.54}$$

which immediately lead to an expression for the Hadamard matrix in terms of the probability amplitudes for quantum entanglement (Duarte and Taylor 2019)

$$H = (|\psi\rangle_+ + |\psi\rangle^-) \tag{F.55}$$

which is equivalent to

$$H = 2^{-1/2}(\sigma_x + \sigma_z) \tag{F.56}$$

References

Ayres F 1965 *Modern Algebra* (New York: McGraw-Hill)
Dirac P A M 1978 *The Principles of Quantum Mechanics* 4th edn (Oxford: Oxford University Press)

Duarte F J, Taylor T S and Slaten J C 2019 Unpublished

Duarte F J and Taylor 2019 Unpublished

Fowles G R 1968 *Introduction to Modern Optics* (New York: Holt, Rinehart, and Winston)

Nielsen M A and Chuang I L 2000 *Quantum Computation and Quantum Information* (Cambridge: Cambridge University Press)

Ortega J M 1987 *Matrix Theory* (New York: Plenum)

Robson B A 1974 *The Theory of Polarization Phenomena* (Oxford: Clarendon)

Zehfuss G 1858 Über eine gewisse determinante *Z. Math. Phys.* **3** 298–301

Appendix G

Trigonometric identities

Trigonometric identities useful in the calculation of probability amplitudes and probabilities related to polarization are listed.

G.I Trigonometric identities

The following are timeless and useful trigonometric identities:

$$\sin^2 \varphi + \cos^2 \varphi = 1 \tag{G.1}$$

$$\sin(-\varphi) = -\sin \varphi \tag{G.2}$$

$$\cos(-\varphi) = \cos \varphi \tag{G.3}$$

$$\sin(\varphi + \theta) = \sin \varphi \cos \theta + \cos \varphi \sin \theta \tag{G.4}$$

$$\sin(\varphi - \theta) = \sin \varphi \cos \theta - \cos \varphi \sin \theta \tag{G.5}$$

$$\cos(\varphi + \theta) = \cos \varphi \cos \theta - \sin \varphi \sin \theta \tag{G.6}$$

$$\cos(\varphi - \theta) = \cos \varphi \cos \theta + \sin \varphi \sin \theta \tag{G.7}$$

$$\sin 2\varphi = 2 \sin \varphi \cos \varphi \tag{G.8}$$

$$\cos 2\varphi = \cos^2 \varphi - \sin^2 \varphi \tag{G.9}$$

$$\cos 2\varphi = 1 - 2 \sin^2 \varphi \tag{G.10}$$

$$\cos 2\varphi = 2 \cos^2 \varphi - 1 \tag{G.11}$$

$$2 \sin^2 \varphi = 1 - \cos 2\varphi \tag{G.12}$$

$$2 \cos^2 \varphi = 1 + \cos 2\varphi \tag{G.13}$$

$$\cos \varphi + \cos \theta = 2 \cos(\varphi - \theta)\cos(\varphi + \theta) \tag{G.14}$$

$$\cos \varphi - \cos \theta = -2 \sin(\varphi - \theta)\sin(\varphi + \theta) \tag{G.15}$$

$$-\cos \varphi + \cos \theta = 2 \sin(\varphi - \theta)\sin(\varphi + \theta) \tag{G.16}$$

$$-\cos \varphi - \cos \theta = -2 \cos(\varphi - \theta)\cos(\varphi + \theta) \tag{G.17}$$

$$\sin \varphi + \sin \theta = 2 \cos(\varphi - \theta)\sin(\varphi + \theta) \tag{G.18}$$

$$\sin \varphi - \sin \theta = 2 \sin(\varphi - \theta)\cos(\varphi + \theta) \tag{G.19}$$

$$-\sin \varphi + \sin \theta = -2 \sin(\varphi - \theta)\cos(\varphi + \theta) \tag{G.20}$$

$$-\sin \varphi - \sin \theta = -2 \cos(\varphi - \theta)\sin(\varphi + \theta) \tag{G.21}$$

$$\sin \varphi \sin \theta = \frac{1}{2}(\cos(\varphi - \theta) - \cos(\varphi + \theta)) \tag{G.22}$$

$$\cos \varphi \cos \theta = \frac{1}{2}(\cos(\varphi - \theta) + \cos(\varphi + \theta)) \tag{G.23}$$

$$\sin \varphi \cos \theta = \frac{1}{2}(\sin(\varphi + \theta) + \sin(\varphi - \theta)) \tag{G.24}$$

$$\cos \varphi \sin \theta = \frac{1}{2}(\sin(\varphi + \theta) - \sin(\varphi - \theta)). \tag{G.25}$$

Of particular interest are the identities related to the Pryce–Ward angle 2φ given in equations (G.8)–(G.13).

Appendix H

More on quantum notation

Further aspects of quantum notation, as per *Dirac's identities*, are explored.

H.1 Introduction

Although it might be obvious to most readers, here the quantumness of equations such as $|\psi\rangle = (|x, y\rangle - |y, x\rangle)$ is examined further.

H.2 Certainly not classical

Already in 1926, Born, Heisenberg, and Pascal had brought to the physics world their amazing discovery succinctly expressed as (Born *et al* 1926)

$$(PQ - QP) = \frac{h}{2\pi i} \tag{H.1}$$

$$(PQ - QP) = -i\hbar \tag{H.2}$$

which is known as the *commutation rule*. This equation, discovered by Heisenberg, Born, and Pascal, illustrates rather dramatically that in the quantum world expressions such as $(AB - BA)$ do not banish, that is,

$$(AB - BA) \neq 0. \tag{H.3}$$

This can be easily demonstrated via the expression

$$(\sigma_x\sigma_y - \sigma_y\sigma_x) \tag{H.4}$$

$$\begin{pmatrix} 0 & 1 \\ 1 & 0 \end{pmatrix}\begin{pmatrix} 0 & -i \\ i & 0 \end{pmatrix} - \begin{pmatrix} 0 & -i \\ i & 0 \end{pmatrix}\begin{pmatrix} 0 & 1 \\ 1 & 0 \end{pmatrix} = \begin{pmatrix} 2i & 0 \\ 0 & -2i \end{pmatrix}. \tag{H.5}$$

Moreover, the *Poisson bracket* is defined as (Dirac 1978)

$$(\hat{H}\hat{A} - \hat{A}\hat{H}) = [\hat{A}, \hat{H}]. \tag{H.6}$$

In this equation, $\hat{H}$ and $\hat{A}$ are operators, and $\hat{H}$ is known as the Hamiltonian (Feynman *et al* 1965).

H.3 Multiplication of probability amplitudes

In chapter III of his book, in a section entitled 'Developments in notation', Dirac introduces the commutative axiom of multiplication associated with *ket* vectors (Dirac 1978). In particular, he introduces the identity

$$|a\rangle|b\rangle = |b\rangle|a\rangle. \tag{H.7}$$

Dirac does so using the preamble, 'We assume that they have a product $|a\rangle|b\rangle$ for which the commutative and distributive axioms of multiplication hold'.

Here, a subtlety arises: Dirac's commutative axiom applies perfectly if $|a\rangle$ and $|b\rangle$ are probability amplitudes represented by complex wave functions. For instance,

$$|a\rangle|b\rangle = \psi_1 e^{-i(\phi_1+\theta_1)}\psi_2 e^{-i(\phi_2+\theta_2)} = \psi_1\psi_2 e^{-i(\phi_1+\theta_1+\phi_2+\theta_2)} \tag{H.8}$$

$$|b\rangle|a\rangle = \psi_2 e^{-i(\phi_2+\theta_2)}\psi_1 e^{-i(\phi_1+\theta_1)} = \psi_2\psi_1 e^{-i(\phi_2+\theta_2+\phi_1+\theta_1)} \tag{H.9}$$

thus clearly showing that $|a\rangle|b\rangle = |b\rangle|a\rangle$.

However, the same commutative axiom does not apply if the *kets* are associated with vectors. For instance, if

$$|x\rangle = \begin{pmatrix} 1 \\ 0 \end{pmatrix} \tag{H.10}$$

$$|y\rangle = \begin{pmatrix} 0 \\ 1 \end{pmatrix} \tag{H.11}$$

using the direct vector product (see chapter 24, and appendix F) yields

$$|x\rangle|y\rangle = \begin{pmatrix} 1 \\ 0 \end{pmatrix} \cdot \begin{pmatrix} 0 \\ 1 \end{pmatrix} = \begin{pmatrix} 0 & 1 \\ 0 & 0 \end{pmatrix} \tag{H.12}$$

and

$$|y\rangle|x\rangle = \begin{pmatrix} 0 \\ 1 \end{pmatrix} \cdot \begin{pmatrix} 1 \\ 0 \end{pmatrix} = \begin{pmatrix} 0 & 0 \\ 1 & 0 \end{pmatrix} \tag{H.13}$$

clearly showing that $|x\rangle|y\rangle \neq |y\rangle|x\rangle$. If instead of the direct vector product the Kronecker $\otimes$ product is utilized,

$$|x\rangle|y\rangle = \begin{pmatrix} 1 \\ 0 \end{pmatrix} \otimes \begin{pmatrix} 0 \\ 1 \end{pmatrix} = \begin{pmatrix} 0 \\ 1 \\ 0 \\ 0 \end{pmatrix} = \begin{pmatrix} 0 & 1 \\ 0 & 0 \end{pmatrix} \qquad \text{(H.14)}$$

$$|y\rangle|x\rangle = \begin{pmatrix} 0 \\ 1 \end{pmatrix} \otimes \begin{pmatrix} 1 \\ 0 \end{pmatrix} = \begin{pmatrix} 0 \\ 0 \\ 1 \\ 0 \end{pmatrix} = \begin{pmatrix} 0 & 0 \\ 1 & 0 \end{pmatrix} \qquad \text{(H.15)}$$

thus confirming the previous result. The conversion from a 4×1 vector to a 2×2 vector is performed via the *vec* function (Neudecker 1969).

Assuming that $|a\rangle|b\rangle = |b\rangle|a\rangle$ applies, then some interesting effects in notation can arise. For instance, since

$$(|x, y\rangle - |y, x\rangle) = (|x\rangle|y\rangle - |y\rangle|x\rangle) \neq 0 \qquad \text{(H.16)}$$

$$(|x\rangle|y\rangle - |x\rangle|y\rangle) \neq 0 \qquad \text{(H.17)}$$

since the second term in parenthesis can be expressed as

$$|x\rangle|y\rangle = |y\rangle|x\rangle \qquad \text{(H.18)}$$

This would establish a complete equivalence between equations (H.17) and (H.16).

References

Born M, Heisenberg W and Jordan P 1926 Zur quantenmechanik II *Z. Phys.* **35** 557–615

Dirac P A M 1978 *The Principles of Quantum Mechanics* 4th edn (Oxford: Oxford University Press)

Feynman R P, Leighton R B and Sands M 1965 *The Feynman Lectures on Physics* vol III (Reading, MA: Addison-Wesley)

Neudecker H 1969 Some theorems on matrix differentiation with special reference to Kronecker matrix products *J. Am. Stat. Assoc.* **64** 953–63

IOP Publishing

Fundamentals of Quantum Entanglement

F J Duarte

Appendix I

From quantum principles to classical optics

The Dirac–Feynman principle is utilized to derive the generalized interference equation that can then be applied to explain interference, diffraction, refraction, and reflection in a succinct, hierarchical, and unified manner. The uncertainty principle and the cavity linewidth equation are also explained in interferometric terms.

I.1 Introduction

In chapter 29, the idea articulated by Lamb (1987), essentially advocating that this is a quantum world, was endorsed via the introduction of the DFL doctrine. Dirac's name was incorporated since he was the first to describe macroscopic interference using quantum principles (Dirac 1978). Feynman's name was also incorporated since he used quantum path integrals to describe 'classical' beam divergence (Feynman and Hibbs 1965). Here it is shown that the generalized quantum interference equation can be applied to describe, in a unified, cohesive, and hierarchical approach, interference, diffraction, refraction, and reflection (Duarte 1997, 2003).

I.2 From quantum interference to generalized diffraction

The Dirac–Feynman probability amplitude

$$\langle x|s \rangle = \sum_{j=1}^{N} \langle x|j \rangle \langle j|s \rangle \tag{I.1}$$

leads, as explained in chapter 2, to the generalized probability in one dimension (Duarte 1991, 1993):

$$\langle x|s \rangle \langle x|s \rangle^* = \left(\sum_{j=1}^{N} \langle x|j \rangle \langle j|s \rangle \right) \left(\sum_{j=1}^{N} \langle x|j \rangle \langle j|s \rangle \right)^* \tag{I.2}$$

doi:10.1088/2053-2563/ab2b33ch38   I-1

$$\langle x|s\rangle\langle x|s\rangle^* = \sum_{j=1}^{N} \Psi(r_j) \sum_{m=1}^{N} \Psi(r_m) e^{i(\Omega_m - \Omega_j)} \tag{I.3}$$

$$\langle x|s\rangle\langle x|s\rangle^* = \sum_{j=1}^{N} \Psi(r_j)^2 + 2 \sum_{j=1}^{N} \Psi(r_j) \left(\sum_{m=j+1}^{N} \Psi(r_m) \cos(\Omega_m - \Omega_j) \right). \tag{I.4}$$

These three equivalent probability equations apply to single-photon propagation or to the propagation of ensembles of indistinguishable photons (Duarte 1993).

From the phase term of equation (I.4), it can be shown that (Duarte 1997, 2006)

$$d_m(\pm n_1 \sin \Theta_m \pm n_2 \sin \Phi_m)\frac{2\pi}{\lambda_v} = M\pi \tag{I.5}$$

where Θ_m and Φ_m are the angles of incidence and diffraction, respectively, n_1 and n_2 are the refractive indices prior- and post-diffraction, and $M = 0, 2, 4, 6, \ldots$.

For $n_1 = n_2$, $\lambda = \lambda_v$, and equation (I.5) reduces to the *generalized diffraction grating equation*

$$d_m(\pm \sin \Theta_m \pm \sin \Phi_m) = m\lambda \tag{I.6}$$

where $m = 0, 1, 2, 3, \ldots$ are the various *diffraction orders*.

I.3 From generalized diffraction to generalized refraction

For the condition $d_m \ll \lambda$, the diffraction grating equation can only be solved for (Duarte 1997)

$$d_m(\pm n_1 \sin \Theta_m \pm n_2 \sin \Phi_m)\frac{2\pi}{\lambda_v} = 0 \tag{I.7}$$

which leads directly to the *generalized refraction equation*

$$(\pm n_1 \sin \Theta_m \pm n_2 \sin \Phi_m) = 0. \tag{I.8}$$

For the case of incidence below the normal (−) and refraction above the normal (+), equation (I.8) becomes (Duarte 2006)

$$-n_1 \sin \Theta_m + n_2 \sin \Phi_m = 0 \tag{I.9}$$

and

$$n_1 \sin \Theta_m = n_2 \sin \Phi_m \tag{I.10}$$

is the well-known *equation of refraction*, also known as *Snell's law*.

For the case of incidence above the normal (+) and refraction above the normal (+) (Duarte 2006),

$$+n_1 \sin \Theta_m + n_2 \sin \Phi_m = 0 \tag{I.11}$$

and

$$n_1 \sin \Theta_m = -n_2 \sin \Phi_m \qquad (I.12)$$

which is the refraction law for *negative refraction*. This subject is treated in greater detail by Duarte (2015).

I.4 From generalized refraction to reflection

From the generalized equation of refraction, equation (I.8) (Duarte 1997, 2015), for $n_1 = n_2$,

$$(\pm \sin \Theta_m \pm \sin \Phi_m) = 0. \qquad (I.13)$$

For incidence above the normal (+) and reflection below the normal (−),

$$+\sin \Theta_m - \sin \Phi_m = 0 \qquad (I.14)$$

which means

$$\Theta_m = \Phi_m \qquad (I.15)$$

where Θ_m is the angle of incidence, and Φ_m is the angle of reflection. This is the well-known *law of reflection*.

I.5 From quantum interference to Heisenberg's uncertainty principle

In this section an approximate geometrical derivation of Heisenberg's uncertainty principle, via the generalized probability equation for interference, is illustrated. As already explained, from the generalized interferometric equation

$$|\langle x|s \rangle|^2 = \sum_{j=1}^{N} \Psi(r_j)^2 + 2 \sum_{j=1}^{N} \Psi(r_j) \left(\sum_{m=j+1}^{N} \Psi(r_m) \cos(\Omega_m - \Omega_j) \right)$$

emerges the generalized diffraction equation (I.6) from which, for positive diffraction, the usual equation of diffraction

$$d_m(\sin \Theta_m \pm \sin \Phi_m) = m\lambda \qquad (I.16)$$

emerges. For $\Theta_m \approx \Phi_m(=\theta)$, the Littrow grating equation

$$2d \sin \theta = m\lambda \qquad (I.17)$$

can be established.

For an expanded beam of light incident on a reflection diffraction grating, as depicted in figure I1, $\sin \theta = \Delta x/l$, where l is the length of the grating, and Δx is the path difference. For an infinitesimal change in wavelength, at two infinitesimally different wavelengths, from equation (I.17)

$$\lambda_1 = \frac{2d}{m} \left(\frac{\Delta x_1}{l} \right) \qquad (I.18)$$

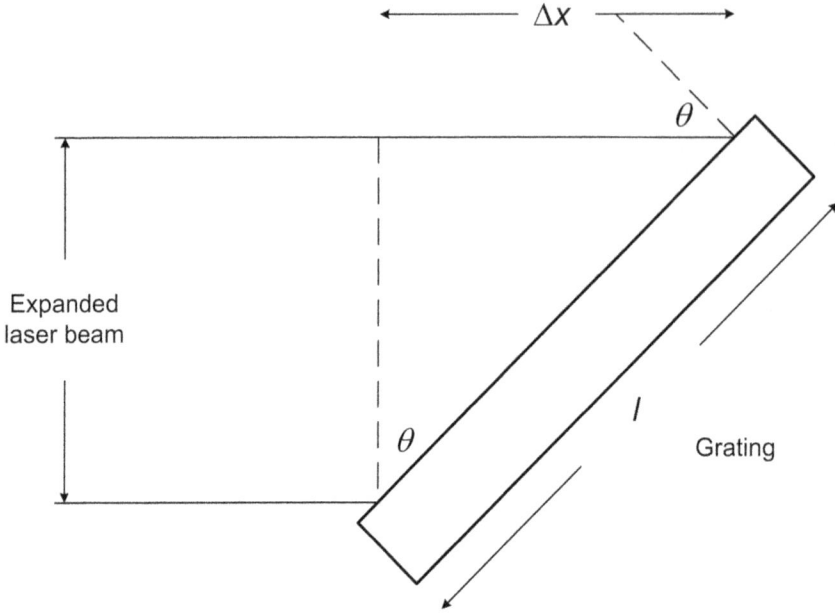

Figure I1. Reflection diffraction grating interacting with expanded laser beam.

$$\lambda_2 = \frac{2d}{m}\left(\frac{\Delta x_2}{l}\right). \tag{I.19}$$

Since equation (I.17) can also be expressed as

$$\frac{2d}{m} = \frac{l\lambda}{\Delta x} \tag{I.20}$$

$\Delta\lambda = (\lambda_1 - \lambda_2)$ yields

$$\Delta\lambda = \frac{l\lambda}{\Delta x}\left(\frac{\Delta x_1 - \Delta x_2}{l}\right). \tag{I.21}$$

To distinguish between a maxima and minima, the *difference* in path differences should be $(\Delta x_1 - \Delta x_2) \approx \lambda$. Thus, equation (I.21) reduces immediately to the diffraction identity

$$\Delta\lambda \approx \frac{\lambda^2}{\Delta x}. \tag{I.22}$$

The momentum expression $p = \hbar k$ for two slightly different wavelengths leads to

$$p_1 - p_2 = \frac{h(\lambda_1 - \lambda_2)}{\lambda_1\lambda_2} \tag{I.23}$$

and by restating the assumption of two infinitesimally different wavelengths, this equation reduces to

$$\Delta p \approx h \frac{\Delta \lambda}{\lambda^2}. \tag{I.24}$$

Substitution of equation (I.22) into (I.24) immediately yields

$$\Delta p \Delta x \approx h \tag{I.25}$$

which is *Heisenberg's Uncertainty Principle* (Dirac 1978). Additional useful forms of the uncertainty principle are its frequency–spatial version

$$\Delta \nu \Delta x \approx c \tag{I.26}$$

and its frequency–time version

$$\Delta \nu \Delta t \approx 1. \tag{I.27}$$

I.6 The cavity linewidth equation

It has already been established that the generalized interferometric equation (I.4) leads to the generalized diffraction equation (I.6) from which, for positive diffraction, the usual equation of diffraction

$$d_m (\pm \sin \Theta_m \pm \sin \Phi_m) = m\lambda \tag{I.28}$$

is obtained. As seen previously, for $\Theta_m \approx \Phi_m (=\theta)$, the Littrow grating equation

$$2d \sin \theta = m\lambda$$

can be established.

Following Duarte (1992) and considering two slightly different wavelengths, an expression for $(\lambda_1 - \lambda_2) = \Delta \lambda$ can be written as

$$\Delta \lambda = \frac{2d}{m} (\sin \theta_1 - \sin \theta_2) \tag{I.29}$$

and for $\theta_1 \approx \theta_2 (=\theta)$ as

$$\Delta \lambda \approx \frac{2d}{m} \Delta \theta \left(1 - \frac{3\theta^2}{3!} + \frac{5\theta^4}{5!} - \frac{7\theta^6}{7!} + \cdots \right). \tag{I.30}$$

Differentiation of the Littrow grating equation yields

$$\frac{\partial \theta}{\partial \lambda} \cos \theta = \frac{m}{2d} \tag{I.31}$$

and substitution of $(m/2d)$ into equation (I.30) leads to

$$\Delta \lambda \approx \Delta \theta \left(\frac{\partial \theta}{\partial \lambda} \right)^{-1} \left(1 - \frac{\theta^2}{2!} + \frac{\theta^4}{4!} - \frac{\theta^6}{6!} \cdots \right) (\cos \theta)^{-1}. \tag{I.32}$$

Since the expansion in θ approaches $\cos\theta$,

$$\Delta\lambda \approx \Delta\theta\left(\frac{\partial\theta}{\partial\lambda}\right)^{-1} \tag{I.33}$$

or

$$\Delta\lambda \approx \Delta\theta(\nabla_\lambda\theta)^{-1} \tag{I.34}$$

which is the well-known cavity linewidth equation (Duarte 1992).

I.7 Generalized multiple-prism dispersion

Since the cavity linewidth equation $\Delta\lambda \approx \Delta\theta(\nabla_\lambda\theta)^{-1}$ depends inversely on the overall cavity dispersion $(\nabla_\lambda\theta)$, it is important to have access to generalized equations of intracavity dispersion applicable to multiple-prism grating assemblies. Precise knowledge of the overall intracavity dispersion is essential for the design of optimized high-power pulsed tunable lasers. For instance, a high peak-power optimized multiple-prism solid-state dye laser oscillator can yield diffraction-limited beam divergence at a linewidth of $\Delta\nu \approx 350$ MHz for a near-Gaussian temporal pulse at $\Delta t \approx 3$ ns, so that $\Delta\nu\Delta t \approx 1.05$, leading to a performance near the frequency–time limit allowed by Heisenberg's uncertainty principle (Duarte 1999).

For a generalized multiple-prism array, as illustrated in figure I2, and the generalized diffraction equation (I.8) (Duarte and Piper 1982, Duarte 2006),

$$\phi_{1,m} + \phi_{2,m} = \varepsilon_m \pm \alpha_m \tag{I.35}$$

$$\psi_{1,m} + \psi_{2,m} = \alpha_m \tag{I.36}$$

$$\sin\phi_{1,m} = \pm n_m \sin\psi_{1,m} \tag{I.37}$$

$$\sin\phi_{2,m} = \pm n_m \sin\psi_{2,m}. \tag{I.38}$$

Here, $\phi_{1,m}$ and $\phi_{2,m}$ are the angles of incidence and emergence, and $\psi_{1,m}$ and $\psi_{2,m}$ are the corresponding angles of refraction at the mth prism. In these equations the positive sign + indicates positive refraction while the negative sign − refers to negative refraction.

Differentiating equations (I.37) and (I.38) and using the derivative identity $(d\psi_{1,m}/dn) = -(d\psi_{2,m}/dn)$ leads to the single-pass dispersion following the mth prism (Duarte and Piper, 1982, 1983, Duarte 2006):

$$\nabla_\lambda\phi_{2,m} = \pm\mathcal{H}_{2,m}\nabla_\lambda n_m \pm (k_{1,m}k_{2,m})^{-1}(\mathcal{H}_{1,m}\nabla_\lambda n_m(\pm)\nabla_\lambda\phi_{2,(m-1)}) \tag{I.39}$$

where $\nabla_\lambda = \partial/\partial\lambda$ and

$$k_{1,m} = \frac{\cos\psi_{1,m}}{\cos\phi_{1,m}} \tag{I.40}$$

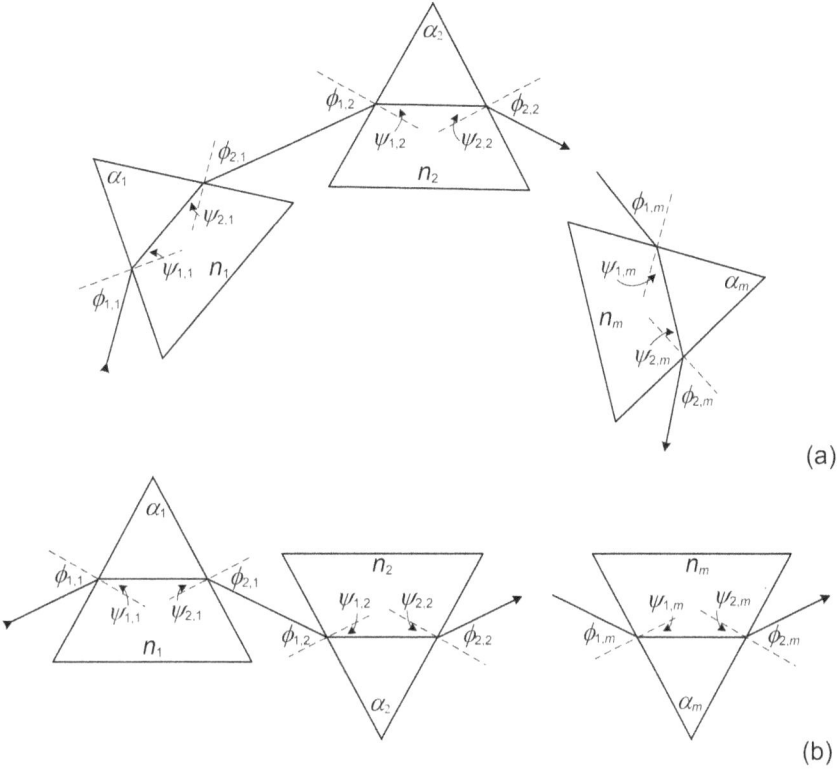

Figure I2. Multiple-prism arrays with the (a) additive configuration and (b) compensating configuration.

$$k_{2,m} = \frac{\cos \phi_{2,m}}{\cos \psi_{2,m}} \tag{I.41}$$

$$\mathcal{H}_{1,m} = \frac{\tan \phi_{1,m}}{n_m} \tag{I.42}$$

$$\mathcal{H}_{2,m} = \frac{\tan \phi_{2,m}}{n_m}. \tag{I.43}$$

$k_{1,m}$ and $k_{2,m}$ represent the beam expansion, at the mth prism, by the incidence and the emergence beams, respectively. In equation (I.39), ($\pm$) refers to deployment in either a positive (+) or compensating (−) configuration, while the simple $\pm$ indicates either positive or negative refraction (Duarte 2006).

Differentiation of equation (I.41) leads to the generalized single-pass dispersion equation for positive refraction

$$\nabla_\lambda \phi_{2,m} = \mathcal{H}_{2,m} \nabla_\lambda n_m + (k_{1,m} k_{2,m})^{-1}(\mathcal{H}_{1,m} \nabla_\lambda n_m \pm \nabla_\lambda \phi_{2,(m-1)}). \tag{I.44}$$

From this equation it can be shown that the generalized *double-pass* multiple-prism dispersion is given by (Duarte 1985, 1989)

$$
\begin{aligned}
\nabla_\lambda \Phi_P = {} & 2M_1 M_2 \sum_{m=1}^{r} (\pm 1) \mathcal{H}_{1,m} \left(\prod_{j=m}^{r} k_{1,j} \prod_{j=m}^{r} k_{2,j} \right)^{-1} \nabla_\lambda n_m \\
& + 2 \sum_{m=1}^{r} (\pm 1) \mathcal{H}_{2,m} \left(\prod_{j=1}^{m} k_{1,j} \prod_{j=1}^{m} k_{2,j} \right) \nabla_\lambda n_m
\end{aligned}
\tag{I.45}
$$

$$
M_1 = \prod_{m=1}^{r} k_{1,m}
\tag{I.46}
$$

$$
M_2 = \prod_{m=1}^{r} k_{2,m}.
\tag{I.47}
$$

Furthermore, the overall multiple-prism grating multi return-pass laser linewidth is given by (Duarte and Piper 1984, Duarte 2001)

$$
\Delta \lambda_R = \Delta \theta_R (RM \nabla_\lambda \Theta_G + R \nabla_\lambda \Phi_P)^{-1}
\tag{I.48}
$$

where R is the number of return intracavity passes elapsed from the leading edge of the laser excitation pulse to the onset of laser emission (Duarte 2001). This equation neatly shows the enormous effect on laser emission linewidth that intracavity multiple-prism factors in the $100 \leqslant M \leqslant 200$ range can have since the grating dispersion is multiplied by M (Duarte 2015).

I.7.1 Generalized multiple-prism dispersion for laser pulse compression

It has been established that the measured laser linewidth $\Delta \lambda \approx \Delta \theta (\nabla_\lambda \theta)^{-1}$ of an optimized multiple-prism grating laser oscillator can be very narrow and even approach the limit imposed by Heisenberg's uncertainty principle. For laser pulse compression, the reverse is desired. That is, since $\Delta \nu \Delta t \approx 1$, a very broadband laser emission can lead to a very narrow temporal pulse, and the least amount of intracavity dispersion is required. This requires detailed knowledge of the first, second, third, and even higher derivatives of the intracavity dispersion.

Using the identity (Duarte 1987),

$$
\nabla_n \phi_{2,m} = \nabla_\lambda \phi_{2,m} (\nabla_\lambda n_m)^{-1}
\tag{I.49}
$$

where $\nabla_n = \partial / \partial n$, the generalized single-pass dispersion, that is, equation (I.44), becomes (Duarte 1987, 2009)

$$
\nabla_n \phi_{2,m} = \mathcal{H}_{2,m} + (\mathcal{M})^{-1} (\mathcal{H}_{1,m} \pm \nabla_n \phi_{2,(m-1)})
\tag{I.50}
$$

where

$$
(\mathcal{M})^{-1} = k_{1,m}^{-1} k_{2,m}^{-1}.
\tag{I.51}
$$

The second derivative of $\phi_{2,m}$, that is, $\nabla_n^2\phi_{2,m}$, is given by (Duarte 1987, 2000)

$$
\begin{aligned}
\nabla_n^2\phi_{2,m} &= \nabla_n\mathcal{H}_{2,m} \\
&+ (\nabla_n\mathcal{M}^{-1})(\mathcal{H}_{1,m} \pm \nabla_n\phi_{2,(m-1)}) \\
&+ (\mathcal{M}^{-1})\left(\nabla_n\mathcal{H}_{1,m} \pm \nabla_n^2\phi_{2,(m-1)}\right).
\end{aligned}
\tag{I.52}
$$

The third derivative of $\phi_{2,m}$, $\nabla_n^3\phi_{2,m}$, is given by (Duarte 2009)

$$
\begin{aligned}
\nabla_n^3\phi_{2,m} &= \nabla_n^2\mathcal{H}_{2,m} \\
&+ \left(\nabla_n^2\mathcal{M}^{-1}\right)(\mathcal{H}_{1,m} \pm \nabla_n\phi_{2,(m-1)}) \\
&+ 2(\nabla_n\mathcal{M}^{-1})\left(\nabla_n\mathcal{H}_{1,m} \pm \nabla_n^2\phi_{2,(m-1)}\right) \\
&+ (\mathcal{M}^{-1})\left(\nabla_n^2\mathcal{H}_{1,m} \pm \nabla_n^3\phi_{2,(m-1)}\right).
\end{aligned}
\tag{I.53}
$$

The fourth derivative of $\phi_{2,m}$, $\nabla_n^4\phi_{2,m}$, is given by (Duarte 2009)

$$
\begin{aligned}
\nabla_n^4\phi_{2,m} &= \nabla_n^3\mathcal{H}_{2,m} \\
&+ \left(\nabla_n^3\mathcal{M}^{-1}\right)(\mathcal{H}_{1,m} \pm \nabla_n\phi_{2,(m-1)}) \\
&+ 3\left(\nabla_n^2\mathcal{M}^{-1}\right)\left(\nabla_n\mathcal{H}_{1,m} \pm \nabla_n^2\phi_{2,(m-1)}\right) \\
&+ 3(\nabla_n\mathcal{M}^{-1})\left(\nabla_n^2\mathcal{H}_{1,m} \pm \nabla_n^3\phi_{2,(m-1)}\right) \\
&+ (\mathcal{M}^{-1})\left(\nabla_n^3\mathcal{H}_{1,m} \pm \nabla_n^4\phi_{2,(m-1)}\right)
\end{aligned}
\tag{I.54}
$$

and so on.

Eventually, the seventh derivative of $\phi_{2,m}$, that is, $\nabla_n^7\phi_{2,m}$, is given by

$$
\begin{aligned}
\nabla_n^7\phi_{2,m} &= \nabla_n^6\mathcal{H}_{2,m} \\
&+ \left(\nabla_n^6\mathcal{M}^{-1}\right)(\mathcal{H}_{1,m} \pm \nabla_n\phi_{2,(m-1)}) \\
&+ 6\left(\nabla_n^5\mathcal{M}^{-1}\right)\left(\nabla_n\mathcal{H}_{1,m} \pm \nabla_n^2\phi_{2,(m-1)}\right) \\
&+ 15\left(\nabla_n^4\mathcal{M}^{-1}\right)\left(\nabla_n^2\mathcal{H}_{1,m} \pm \nabla_n^3\phi_{2,(m-1)}\right) \\
&+ 20\left(\nabla_n^3\mathcal{M}^{-1}\right)\left(\nabla_n^3\mathcal{H}_{1,m} \pm \nabla_n^4\phi_{2,(m-1)}\right) \\
&+ 15\left(\nabla_n^2\mathcal{M}^{-1}\right)\left(\nabla_n^4\mathcal{H}_{1,m} \pm \nabla_n^5\phi_{2,(m-1)}\right) \\
&+ 6(\nabla_n\mathcal{M}^{-1})\left(\nabla_n^5\mathcal{H}_{1,m} \pm \nabla_n^6\phi_{2,(m-1)}\right) \\
&+ (\mathcal{M}^{-1})\left(\nabla_n^6\mathcal{H}_{1,m} \pm \nabla_n^7\phi_{2,(m-1)}\right).
\end{aligned}
\tag{I.55}
$$

For this series of derivatives, the numerical factors can be predetermined from Pascal's triangle relative to N, where $(N + 1)$ is the order of the derivative (Duarte 2009, 2018).

Observing the series of derivatives, $\nabla_n^1 \phi_{2,m}, \nabla_n^2 \phi_{2,m}, \nabla_n^3 \phi_{2,m} \ldots \nabla_n^7 \phi_{2,m}$, a generalized equation for the higher derivatives is found (Duarte 2013):

$$\nabla_n^r \phi_{2,m} = \nabla_n^{r-1} \mathcal{H}_{2,m} + (\mathcal{M})^{-1} (\nabla_n + \zeta)^{r-1} \tag{I.56}$$

where

$$\zeta^s = \nabla_n^s \mathcal{H}_{1,m} \pm \nabla_n^{s+1} \phi_{2,(m-1)} \tag{I.57}$$

$$\zeta^0 = 1 = \mathcal{H}_{1,m} \pm \nabla_n \phi_{2,(m-1)}. \tag{I.58}$$

The lower derivatives have been used to design multiple-prism pulse compressors, including up to six prisms, for semiconductor lasers (Pang *et al* 1992). The lower derivatives have also been used by Osvay *et al* (2004, 2005) in practical femtosecond lasers to calculate intracavity dispersions and laser pulse durations for double-prism compressors, finding good agreement between theory and experiments.

I.8 Discussion

Here it was clearly and unambiguously demonstrated that from a purely quantum equation

$$\langle x|s \rangle \langle x|s \rangle^* = \left(\sum_{j=1}^{N} \langle x|j \rangle \langle j|s \rangle \right) \left(\sum_{j=1}^{N} \langle x|j \rangle \langle j|s \rangle \right)^*$$

and representing the probability amplitudes with complex wave equations, as encouraged by Dirac, the equations for generalized diffraction, generalized refraction, and reflection can be arrived at in a coherent and unified manner. The traditional way to present these equations, via classical optics, is in reverse and in the absence of cohesiveness.

It was also shown that Heisenberg's uncertainty principle can be derived from quantum interferometric principles, thus again demonstrating the enormous significance of interference at the foundations of quantum optics.

Other equations long thought to be entirely classical in nature and origin can be traced back to quantum interferometric principles. Such is the case for the cavity linewidth equation and multiple-prism grating dispersion, a subject that has been of interest since the times of Newton (1704). These are further examples that reinforce the DFL doctrine: the foundation of optics is quantum, and as Willis Lamb implied … it is time we 'learn to enjoy it' (Lamb 1987).

References

Dirac P A M 1978 *The Principles of Quantum Mechanics* 4th edn (Oxford: Oxford University Press)

Duarte F J 1985 Note on achromatic multiple-prism beam expanders *Opt. Commun.* **53** 259–62

Duarte F J 1987 Generalized multiple-prism dispersion theory for pulse compression in ultrafast dye lasers *Opt. Quantum Electron.* **19** 223–9

Duarte F J 1989 Transmission efficiency in achromatic nonorthogonal multiple-prism laser beam expanders *Opt. Commun.* **71** 1–5

Duarte F J 1991 Dispersive dye lasers *High Power Dye Lasers* ed F J Duarte (Berlin: Springer) ch 2

Duarte F J 1992 Cavity dispersion equation $\Delta\lambda \approx \Delta\theta\,(\partial\theta/\partial\lambda)^{-1}$: a note on its origin *Appl. Opt.* **31** 6979–82

Duarte F J 1993 On a generalized interference equation and interferometric measurements *Opt. Commun.* **103** 8–14

Duarte F J 1999 Multiple-prism grating solid-state dye laser oscillator: optimized architecture *Appl. Opt.* **38** 6347–9

Duarte F J 2000 Multiple-prism arrays in laser optics *Am. J. Phys.* **68** 162–6

Duarte F J 2001 Multiple-return-pass beam divergence and the linewidth equation *Appl. Opt.* **40** 3038–41

Duarte F J 2003 *Tunable Laser Optics* (New York: Elsevier)

Duarte F J 2006 Multiple-prism dispersion equations for positive and negative refraction *Appl. Phys.* B **82** 35–8

Duarte F J 1997 Interference, diffraction, and refraction, via Dirac's notation *Am. J. Phys.* **65** 637–40

Duarte F J 2009 Generalized multiple-prism dispersion theory for laser pulse compression: higher order phase derivatives *Appl. Phys.* B **96** 809–14

Duarte F J 2013 Tunable laser optics: applications to optics and quantum optics *Prog. Quantum Electron.* **37** 326–47

Duarte F J 2015 *Tunable Laser Optics* 2nd edn (New York: CRC)

Duarte F J 2018 Mathematical-physics of tunable narrow-linewidth organic laser oscillators *Organic Lasers and Organic Photonics* ed F J Duarte (Bristol: Institute of Physics)

Duarte F J and Piper J A 1982 Dispersion theory of multiple-prism beam expander for pulsed dye lasers *Opt. Commun.* **43** 303–7

Duarte F J and Piper J A 1983 Generalized prism dispersion theory *Am. J. Phys.* **51** 1132–4

Duarte F J and Piper J A 1984 Multi-pass dispersion theory of prismatic pulsed dye lasers *Optica Acta* **31** 331–5

Feynman R P and Hibbs A R 1965 *Quantum Mechanics and Path Integrals* (New York: McGraw-Hill)

Lamb W E 1987 *Schrödingers's cat Paul Adrien Maurice Dirac* ed B N Kursunoglu and E P Wigner (Cambridge: Cambridge University Press) ch 21

Newton I 1704 *Opticks* (London: Royal Society)

Osvay K, Kovács A P, Heiner Z, Kurdi G, Klebniczki J and Csatári M 2004 Angular dispersion and temporal change of femtosecond pulses from misaligned pulse compressors IEEE *J. Select. Top. Quantum Electron.* **10** 213–20

Osvay K, Kovács A P, Kurdi G, Heiner Z, Divall M, Klebniczki J and Ferincz I E 2005 Measurement of non-compensated angular dispersion and the subsequent temporal lengthening of femtosecond pulses in a CPA laser *Opt. Commun.* **248** 201–9

Pang L Y, Fujimoto J G and Kintzer E S 1992 Ultrashort-pulse generation from high-power diode arrays by using intracavity optical nonlinearities *Opt. Lett.* **17** 1599–601

Appendix J

Introduction to Hamilton's quaternions

Here an introduction to Hamilton's quaternions, as utilized in the development of the probability amplitudes for $n = N = 3, 6$, is given.

J.1 Introduction

Hamilton's quaternions were introduced by the mathematician of the same name around the mid-1800s (Hamilton 1866). A more recent review of this subject is given by Koecher and Remmert (1991).

J.2 Basic quaternion identities

Quaternions extend beyond the realm of complex numbers and obey the main relation

$$i^2 = j^2 = k^2 = ijk = -1 \tag{J.1}$$

and the basis elements i, j, and k obey the commutative law when multiplied by 1:

$$i \times 1 = 1 \times i = i \tag{J.2}$$

$$j \times 1 = 1 \times j = j \tag{J.3}$$

$$k \times 1 = 1 \times k = k. \tag{J.4}$$

The self-consistency of the main relation given in equation (J.1) also implies that

$$ij = k \tag{J.5}$$

$$ji = -k \tag{J.6}$$

$$jk = i \tag{J.7}$$

$$kj = -i \tag{J.8}$$

$$ki = j \tag{J.9}$$

$$ik = -j. \tag{J.10}$$

References

Hamilton W R 1866 *Elements of Quaternions* (London: Longman Green & Co)
Koecher M and Remmert R 1991 *Hamilton's quaternions Numbers* (Berlin: Springer) ch 7

Index

www.ingramcontent.com/pod-product-compliance
Lightning Source LLC
Chambersburg PA
CBHW080530220326
41599CB00032B/6258